Herbert Davies, Arthur Templar Davies

The Mechanism of the Circulation of the Blood

Through Organically Diseased Hearts

Herbert Davies, Arthur Templar Davies

The Mechanism of the Circulation of the Blood
Through Organically Diseased Hearts

ISBN/EAN: 9783337211592

Printed in Europe, USA, Canada, Australia, Japan

Cover: Foto ©berggeist007 / pixelio.de

More available books at **www.hansebooks.com**

THE MECHANISM

OF THE

CIRCULATION OF THE BLOOD

THROUGH

ORGANICALLY DISEASED HEARTS

BY

HERBERT DAVIES, M.D., F.R.C.P.

LATE CONSULTING PHYSICIAN TO THE LONDON HOSPITAL, AND FORMERLY
FELLOW OF QUEEN'S COLLEGE, CAMBRIDGE.

EDITED BY

ARTHUR TEMPLER DAVIES

B.A. (NAT. SCIENCE HONOURS), M.B. (CANTAB.), M.R.C.P.

PHYSICIAN TO THE ROYAL HOSPITAL FOR DISEASES OF THE CHEST;
EXAMINING PHYSICIAN TO THE ROYAL NATIONAL HOSPITAL FOR
CONSUMPTION, VENTNOR; ASSISTANT PHYSICIAN TO THE
METROPOLITAN HOSPITAL; LATE CASUALTY PHY-
SICIAN TO ST. BARTHOLOMEW'S HOSPITAL

LONDON
H. K. LEWIS, 136 GOWER STREET, W.C.
1889

EDITOR'S PREFACE.

In editing these papers by my Father, the late Dr. Herbert Davies, a sad and melancholy pleasure falls to my lot. It was his earnest wish and intention to have published them himself during his life-time, but such unfortunately was not to be the case, and shortly before his death, he instructed me to edit them on his behalf. They are the result of many years careful study and matured thought on a subject in which he took the most profound and deep interest. Only those who were intimately acquainted with my Father are aware of the infinite care and trouble which he bestowed on his writings. He never drew any conclusions without most carefully and assiduously verifying the absolute correctness and soundness of his facts. It will be seen that these papers contain several original and new ideas on a subject the mechanical aspect of which was by no means exhausted, and which has important bearings on physiological and pathological problems; but I feel it would be out of place to dwell any further upon the merits of these papers, as they are fully able to speak for themselves.

Through the courtesy and kindness of the Rev. S. Haughton, M.D., F.R.S., Trinity College, Dublin, I am enabled to publish his important mathematical paper on Aortic Regurgitation, which he undertook at my Father's request. I cannot better conclude this preface than in the following lines, which admirably express my own feelings towards the Author of this work.

> " For in my mind
> Is fixed, and now strikes full upon my heart,
> The dear, benign, paternal image, such
> As thine was, when so lately thou didst teach me.
> *   *   *   *   *
> And how I prized the lesson, it behoves
> That long as life endures, my tongue should speak."*

ARTHUR T. DAVIES.

23 Finsbury Square.
May, 1889.

* Dante, *Inferno*, Canto xv.   Cary's Translation.

CONTENTS.

MECHANISM OF THE CIRCULATION

ORGANICALLY DISEASED HEARTS.

It must, no doubt, have often occurred to the minds of many practitioners, when examining a patient, the subject of organic disease of the heart, to have been surprised at the almost unembarrased manner with which the circulation is maintained, in an organ manifestly injured. Perhaps no more striking example of this fact could be found, than in a case recorded by Niemeyer in his "Text-book of Practical Medicine," vol. i., p. 347. A huntsman, in Greifswald, who suffered from extensive stenosis, and insufficiency of the aortic valves, and had immense eccentric hypertrophy of the left ventricle, performed all the manœuvres and forced marches of the army without difficulty. Cases such as this and others come frequently before us, of severe cardiac lesions of long standing, in which no dropsy has ever supervened, and where even the functions

B

of life appear to be carried on with but little inconvenience to the individual. The records of Pathology will furnish very many instances of death, revealing the presence of lesions of the heart, the amount, and even the existence of which, had been unsuspected during life. It will therefore be not only interesting, but practically useful to consider the nature of such cases, and to discuss the mode in which the circulation of the blood is carried on in hearts, which are undoubtedly structurally unsound.

At the commencement of this enquiry, I think it would be advisable to dwell for a short time upon the mechanism of the circulation in its healthy state. I need scarcely say, that in reality we have two hearts, each perfect in itself, each administering to its own share of the circulation, each discharging completely its own proper function, but both intimately and inextricably associated with each other, structurally and functionally by muscular fibres, and inter-communicating nerve ganglia. Two functionally distinct and perfect muscular organs, provided with their respective valvular appliances, and performing very unequal physical tasks, are compelled by co-operative association, to beat and act in unison, and as a necessary consequence of such association to obey the two following fundamental laws :—

1. The corresponding chambers of the right and left hearts receive and expel their contents exactly synchronously, respectively — the two

auricles contracting in unison, and dilating in unison, and the two ventricles exhibiting, respectively, synchronous action in their systole, as well as diastole, and—

2. Equal volumes of blood pass, in equal and the same times, through any two corresponding orifices of the heart.

I will not detail at length the proofs of the truth of these laws, as I have discussed this part of our subject elsewhere, in a paper published in *Proceedings of the Royal Society;* I will here briefly glance at them.

That the ventricles act synchronously is shewn by :—

1. The sphygmographic tracings obtained by M. Marey in his experiments upon the horse. Instruments inserted simultaneously into the right and left chambers, and connected with the recording apparatus, exhibited a complete uniformity in the action of the ventricles, during systole and diastole respectively.

2. Grasping in the hand the exposed heart of a living animal, whether auricles or ventricles, confirms the fact, that the smallest appreciable period of time does not exist, between the respective contractions and dilatations of the corresponding cavities.

* "On the Law which regulates the Relative Magnitude of the Areas of the four Orifices of the Heart," *Proc. Roy. Soc.*, 1870.

3. The auscultator, in listening to the heart, detects only two sounds, although each ventricle produces its first and second sound. The corresponding sounds of each ventricle are exactly blended, and fused into one another, and the four, are, by the synchronous action of the two chambers, reduced to two sounds.

4. Bands of muscular fibres pass from auricle to auricle, and from ventricle to ventricle, compelling thereby the synchronous action of the corresponding chambers.

That equal volumes of blood must traverse, in the same time, the corresponding orifices and chambers of the heart, is shewn by the following considerations.

1. Should the left heart receive, continuously, at every beat from the right heart, more blood (however small the excess) than it sends forth, pulmonary congestion and stagnation must eventually ensue, and the circulation through the entire organ become inevitably arrested, and the supposition of the left heart receiving, at every beat, less blood (however small the deficiency) than it expels subsequently in systole, involves a contradiction, which requires no comment. "There are reasons for believing," say Quain and Sharpey,❋ "that during life, any difference in the capacities of the ventricles is very slight, if it exists at all," and I think we may

* Quain's *Anatomy*, vol. ii., p. 501.

proceed even further and agree with Valentin[*] in his statement, that all the four cavities of the heart, *probably* contain the same quantity of blood during life. " For as the synchronously acting ventricles, expel during their complete systole, the equal volumes of blood, which they had respectively previously received, during their diastole from the auricles, it follows, that the capacities of the auricles must closely equal the capacities of the ventricles, and that all four chambers of the heart must within certain limits, (to be subsequently considered) contain during life, the same quantity of blood." Having established the existence of these two laws, viz., synchroneity of action and equality of received and discharged volumes of blood, and knowing, that the two hearts associated into one organ have very unequal tasks to perform, we may sum up the mechanism of the circulation of the blood, in the healthy heart, in the following propositions :—

The two ventricles, acting with *unequal* forces, send forth equal quantities of blood, synchronously, into the Pulmonic and Systemic circulations respectively, and whatever may be the velocities of the issuing currents, it is evident from the volumes being equal, and their times of flow being the same, that the areas of the pulmonic and aortic

Valentin's *Physiology*, p. 117.

orifices must be inversely, as the velocities of the currents, which traverse them.

2. The two ventricles must receive equal volumes of blood in diastole, but in such a manner, that the momentum of pressure of each in-coming current, is proportional to the inertia of the ventricle, which it has to overcome. By the time, the ventricles are completely filled, all blood movement within them is momentarily suspended, previous to their contents being launched forth into the pulmonary artery and aorta, in directions very different from those, in which the molecules of blood entered the chambers. The areas of the tricuspid and mitral orifices are also, of course, inversely as the velocities of the in-coming currents.

3. The force of each ventricle is such, that the currents through the lungs and general system must be completely, and thoroughly maintained.

These objects must be secured within certain limits of deviation, the large veins and the right auricle on the venous side, and the vessels of the lungs and left auricle on the arterial side becoming compensatory receptacula, for the heart is not a rigid instrument, and although of course always full, does not at all times contain the same quantity of blood, but accommodates itself to the ever varying conditions of the circulation, which arise from muscular pressure and other causes.

Lastly, from the measurements made by Dr.

Peacock and others, the important law* is deduced, that the areas of the four chief apertures of the heart are in proportion, so that whatever ratio the area of the tricuspid bears to that of the mitral, the same ratio does the area of the pulmonic bear to that of the aortic orifice, and as a corollary from this law, and also as the result of actual inspection of the measurement of the orifices, we find that—

$$\frac{\text{The area of the tricuspid}}{\text{The area of the mitral}} = 1\cdot3 \text{ to } 1\cdot4.$$

$$\frac{\text{The area of the pulmonic}}{\text{The area of the aortic}} = 1\cdot3 \text{ to } 1\cdot4.$$

whence the area of any orifice, in either side of the heart, can be determined, when the area of the corresponding orifice on the other side is known.

I have said, that the heart is not a rigid instrument, but allows of deviation in its action, within due limits, according to the circumstances to which it is exposed. The mechanism of the circulation in a person "out of breath" from excessive exertion, will afford the simplest illustration of what I mean. As long as the right and left ventricle are synchronously sending forth equal quantities of blood, into the pulmonic and aortic systems respectively, so long evidently is the left heart—beat per beat—withdrawing from the lungs the same

* *Proceedings of the Royal Society*, No. 118, 1870, Herbert Davies, M.D.

amount of blood, which has been forced into them —beat by beat—by the right heart, and so long manifestly is the quantity of air supplied by the act of respiration, exactly equal to the demand for it. The cardiac and respiratory functions exactly respond to each other—the blood is completely aerated. The two hearts · work smoothly and uniformly in unison, and although the movements of the heart and respiratory muscles may be accelerated by the exercise, "dyspnœa," the cry of unaerated blood, does not occur. This co-ordination of the three muscular forces of the right ventricle, respiratory muscles and left ventricle, is an admirable example of the manner, in which, the animal machine adapts itself irrespective, and even in spite of the volition of the individual, to the varying circumstances of the moment. Increased exercise will cause the venous blood to move onwards with greater velocity and momentum to the right heart, the right auricle and ventricle will be successively more rapidly filled, and more forcibly distended, and therefore stimulated to contract more rapidly per minute, but the correlation of the three forces will at once cause a correspondingly increased rapidity, and more complete development of the respiratory efforts, and a correspondingly quicker filling and emptying of the left heart, and thus, and within due limits, will this increased activity of the cardiac and pulmonic circulation be maintained without distress of breathing, "dyspnœa," taking place.

If, however, by reason of unduly prolonged exertion on the part of the individual, the heart be stimulated to contract at a speed, which the respiratory muscles cannot correspondingly maintain, the period of time allowed for the introduction of air into the bronchial channels and air vesicles, and also for the proper chemical change of each charge of blood, being also insufficient, a portion of each charge from the right ventricle will remain for a time unaerated, and retained in the pulmonary vessels, and possibly in the right ventricle itself.

The quantity of blood thus retained may be in the first instance very small—a few drops or a drachm perhaps—and the two volumes of blood flowing synchronously from the two ventricles will be nearly equal; but if the extreme muscular efforts of the individual be continued, the residue excess of venous blood in the pulmonary vessels will increase at every beat, to such an amount by reason of the excessive functional energy of the right heart, as to produce painful dyspnœa, compelling the individual to moderate or even to arrest at once his bodily exertions. (There is, I suppose, in health an amount of oxygen present in the blood, more than requisite for the immediate wants of the system a reserve amount upon which the individual may

draw, when the body is brought into more power-
ful action. The reserve amount probably varies
in individuals, and accounts for the varying times,
in which they can maintain muscular efforts). At
this period of breathlessness, the effluxes from the
two ventricles are *unequal.* Anyone, who has
watched the countenance of the runners, in a long
and distressing foot-race, cannot have failed to
observe the pale and bluish tinge of the face, indi-
cating the small amount of arterial blood leaving
the left heart, and the gorged conditions of the
right heart. The following case exemplifies the
dangers to which professional trainers are liable.
The Report of it is found in the *Lancet* of May
22nd, 1869. R. M., æt. 50, was admitted into the
Sheffield Public Hospital, under the care of Dr. J.
C. Hall on February 3rd, 1869 with the symptoms
of urgent dyspnœa. He stated, that he had for
many years been a trainer, and a professional
man, and had on one occasion walked a thousand
miles in a thousand consecutive hours (nearly
forty-three days and nights). He was a short
muscular man, and complained of great pain in the
præcordia, and great difficulty of breathing, increased by the slightest exertion. There was
slight prominence of the cardiac region, and the
heart's rhythm was irregular. The first sound
was unnaturally clear at the apex, and the second
sound immediately followed by a slight murmur

heard over the course of the aorta. There was some œdema of the lungs. On February 28th he coughed up about three pints of blood, and died in a few minutes. P. M., *right side* of the heart *much dilated* and the *fibres apparently softened*, the left ventricle slightly hypertrophied. At the back and upper part of the arch of the aorta, and pressing on the trachea, was an aneurysmal tumour of the size of a hen's egg. Everything else was healthy. The frequent and excessive pressure and over-filling, to which the right heart had been exposed, had led to a great dilatation and degeneration of that organ. The aneurysm had had but little or no share in the production of the dilatation, as was shewn in the very slight deviation, from the normal of the cavity of the left ventricle, but its origin lay, no doubt, in the over-strain of the aortic walls, which the life of a professional runner inevitably necessitated. In the above account, I have re-ferred the sense of breathlessness to the presence of an excess of venous blood in the lungs—to a degree of venous stasis in those organs. It would, however, have been much more correct to have ascribed the dyspnœa to a deficiency of oxygen in the system, the feeling of distress being implanted by nature in the chest, in order that the defective supply of oxygen to the tissues, and organs of the body, might be remedied by the act of inspiration, just as the want of food and fluid are felt in the stomach and fauces respectively, as hunger and thirst, for the purpose of pointing out instinctively the paths, by which those necessities are to be supplied. Physiologists tell us, that about twenty cubic inches of air enter the chest at every ordinary inspiration, and that five per cent. of this volume of air, or one cubic inch of oxygen is absorbed and conveyed into the general circulation. If then we consider, that the heart fills and empties itself, about four times during each respiratory act, it is clear that about one-fourth of a cubic inch of oxygen passes through the aortic orifice at each ventricular systole, and we may say, that, during the ordinary condition of the body, the general system *breathes* oxygen through the aortic opening, at the rate of about one-fourth of a cubic inch per beat. The necessity of such a constant supply is evident, when we learn from Bernard's experiment, that the blood of an animal contains as much as ten per cent. of oxygen during fasting,* and from eighteen to twenty per cent. during full digestion, an amount of gas in the arteries, which almost justifies the origin of their name as " air-carrying " vessels.

As the oxygen appears to be mainly conveyed in the blood stream by means of the red corpuscles, and to be incorporated with them of course in a concrete state, it follows, that the perfection of the circulation will essentially depend, upon these multitudinous

* Bernard. Lectures given at the College of France, 1861.

bodies being normal in their amount, and sound in their structure. Should these bodies then become markedly deficient in number, or impaired in their structure or function, dyspnœa must result in consequence of an abnormal supply of oxygen carriers. Hence, breathlessness is a characteristic feature in the pale, chlorotic, anæmic girl, in the sufferer from chronic albuminuria, and in all individuals, whose blood presents deficient means of conveying oxygen into the recesses of the body.

In all such cases, the aorta *breathes* too small an amount of oxygen, and dyspnœa of course supervenes on the most ordinary exertion. These observations point therefore, to the practical importance of maintaining in all cases, the healthy standard of the blood, and will explain, how many individuals pass through many years of their life, exhibiting but little distress from undoubted cardiac mischief, while others, with the same or even a smaller amount of valvular lesion will rapidly sink under their malady. The former individuals, having good assimilating organs, possess an ample amount of red corpuscles ready to supply the tissues with oxygen, even under the difficulties caused by heart disease, while the latter class wanting their due proportion of blood discs, these essentials to existence, rapidly exhaust in the smallest bodily efforts the supply present in the blood, and speedily becoming bankrupt of oxygen, pass into distressing dyspnœa. The causes of dyspnœa are of course very many, but they all ultimately lead to the want of circulating oxygen, and our aim in practice therefore is, whilst removing, if possible, the causes which mechanically obstruct the course of the blood stream, to maintain the normal condition of the liquor sanguinis and blood corpuscles, which play such an important part in the aëration of the body. With normal blood, the compensatory hypertrophy can be developed, and the due amount of oxygen carriers can be maintained, and either failing, the patient succumbs under his malady.

It is clear, therefore, that the gravity of any case of heart disease cannot be measured by the physical signs, alone, which mark its existence, without at the same time taking into full consideration, the state and working of the other functions of the body, the most important element for consideration being the activity and completeness of the blood-making organs, and the perfect action of every portion of the respiratory system. The heart itself has, we shall see, great powers of self-adjustment, which enable it by various degrees of dilatation and hypertrophy, to overcome to a great extent the injurious effects, which would otherwise result from organic changes in its valves and orifices, and no greater mistake can be made in practice, than to decide upon the gravity of a case, in which heart disease exists, upon the examination alone of the organ itself.

MITRAL REGURGITATION.

THE mode in which the circulation of the blood is maintained, in the case of structurally diseased openings, is well shewn by the mechanism exhibited in this instance. Examples are sufficiently numerous, and readily occur to the mind of the practitioner, where mitral regurgitation has existed for many years, without any marked impairment of the function of the heart ; here, as in other instances already quoted of diseased hearts, the compensatory powers of nature overcome for a time the injurious effects, which the morbid lesions produce, and cause the blood to flow evenly as in health, from the right and left ventricle. In other words, the conditions which we have insisted upon so frequently are satisfied namely :—

(*a*) The synchroneity of action of the two ventricles.

(*b*) The equality of effluxes from the right and left ventricle respectively.

In the healthy heart, the left ventricle sends out at each contraction three ounces of blood through the aortic opening. Let us, however, suppose that by reason of imperfect action of the mitral valves, a small quantity (say one drachm) finds its way, in consequence of the ventricular contraction, into the left auricle, meeting the blood, which is already flowing into that chamber; the amount of blood, which is therefore present in the auricle at the close of the ventricular contraction, must be equal to three ounces plus the drachm above mentioned. Ventricular dilatation now takes place, admitting into that chamber the whole of the contents of the auricle, by reason of the natural dilatability within certain limits, of the walls of the ventricle. As the ventricle must now contract upon a larger mass of blood, its functional activity must be increased, in order, that it may completely empty itself, and by this contraction the normal amount (three ounces) is expelled through the aorta, and the drachm regurgitated, as in the previous beat of the heart, into the auricular cavity; this process is continually repeated in exactly similar manner—beat by beat—with the ultimate result of the walls of the ventricle becoming hypertrophied to an amount, corresponding to the increased functional activity, and thus the individual may live for years, exhibiting in his heart a marked mitral regurgitant murmur, but in his daily life, little or no functional disturbance of its action. The compensation is complete, and the diseased heart acts in all respects as a healthy one. It must be borne in mind, that the left auricle becomes also strengthened by the hypertrophy of its walls, as it has to contract each time, after emptying its contents, upon three ounces and a drachm of blood, in place of the normal three ounces; the increased functional energy necessitated, develops the increased muscularity of its parietes. Let us suppose, that the insufficiency of the mitral curtains increases, allowing two drachms of blood to regurgitate into the auricle; as already explained, the three ounces plus two drachms, or the three and a quarter ounces of blood contained in the auricle must dilate the left ventricle to a capacity sufficient to hold it. The ventricle may have sufficient power to expel the whole of its contents, the quarter of an ounce into the auricle, and the three ounces into the aorta, and by the frequent repetition of this process, at every beat, hypertrophy of the ventricle, corresponding to the functional activity required, takes place, and the chamber of the ventricle having become abnormally increased in capacity, by the amount of blood regurgitated, namely, quarter of an ounce, is therefore permanently dilated. Enlargement of the ventricular cavity to hold the extra amount regurgitated, plus the normal amount received from the right side of the heart, is the *first part* of the process, whilst hypertrophy lends its aid to enable the ventricle to expel its contents in two opposite directions. In the process of time, further enlargement of the abnormal opening may occur ; greater dilatation

c

of the ventricular wall will ensue, increased hypertrophy will be developed, and the circulation will be maintained, the contents of the right heart will be evenly conveyed into the systemic circulation, provided that the nutrition of the nerve ganglia and muscular tissue of the heart are maintained. Dilatation of the ventricle and enlargement of the left chamber are not to be considered as a diseased condition, but as being absolutely essential to the circulation of the blood, as without them there would be a stoppage in the passage of the blood from the right to the left side of the heart, and ultimately in the whole venous system. Let us suppose, for example, the left ventricle to be undilatable, beyond the amount of holding the normal three ounces of blood, and the mitral insufficiency to allow of the regurgitation of a quarter of an ounce of that fluid. The ventricular contraction would cause the accumulation of three and a quarter ounces of blood in the auricle, but as in our supposition, the left ventricle can only hold three ounces, a quarter of an ounce must remain behind in the auricle, as the result of that beat. At the next beat the process is repeated, a quarter of an ounce being again regurgitated, and therefore added to the previous quarter of an ounce, and thus in a dozen beats, three ounces of blood will have failed to find its way into the systemic circulation.

Although the pulmonary capillaries act for a time as reservoirs, to hold some of the regurgitant blood, it is clear that such a process could not continue for any length of time, but that the result must inevitably be stagnation of the venous portion of the blood, and death. This *reductio ad absurdum* shows, therefore, that ventricular dilatation must occur to allow of the due mechanism of the circulation, and that it and also its associated hypertrophy are not examples of disease, but on the contrary, are instances of the instinctive conservative power of nature to neutralise the injurious consequences of diseased apertures. It is evident, from what has been stated above, that we may draw the following conclusions, as to the condition of the heart.

1. *Left auricle.*—Temporarily dilated to hold three and a quarter ounces of blood.

   Permanently hypertrophied to expel three and a quarter ounces of blood, instead of three ounces.

2. *Left ventricle.*—Temporarily dilated to hold three and a quarter ounces of blood.

   Permanently hypertrophied to act upon three and a quarter ounces.

3. If (1) and (2) are true, then it also follows, that so long as the left auricle and ventricle completely empty themselves, that—

C 2

(a) The dilatation and hypertrophy of the left auricle and ventricle, increase *pari passu* with the gradual enlargement of the diseased opening, or, in other words, that each chamber is temporarily dilated, and permanently hypertrophied by exactly the amount regurgitated.

(b) That the blood regurgitated is practically a certain amount subtracted from the normal quantity, which passes into the aorta, and therefore by supposing more and more to regurgitate, we subtract more and more from that, which otherwise would have passed into the aorta, and add it to the auricular contents.

It must be also borne in mind, that as the onward movement of the blood from the left auricle, mainly depends on the contractile energy of the right ventricle, the increased volume resulting from the regurgitation of the left ventricle, must call forth increased power from the right ventricle and hypertrophy of the wall of that chamber, and at this stage there will be no dilation of that chamber.

We have thus far considered the case, in which the left ventricle completely empties itself at each systole, the force of the ventricular walls being completely sufficient to the task of emptying the chamber. This task consists—

1. In overcoming at the aortic orifice, the strong resistance of the aortic tension, and the forcible

expulsion of three ounces of blood into that vessel.

2. The expulsion of regurgitant blood through the abnormal mitral opening, in face of the pressure of blood contained in the auricle.

Now the resistance to be overcome at the aortic orifice may be represented by the pressure of a column of blood ninety inches in height, and the resistance at the mitral opening by the pressure of a column of blood say six inches in height. If therefore the ventricular wall fails in vigour, either from general constitutional weakness, or from some incipient degenerative changes in it, its diminished action will be probably most palpable at the aortic orifice, and consequently a smaller amount of blood than usual will find its way for a time into the systemic circulation; at the same time, the amount of regurgitant blood will be but little altered, as the resistance at this opening is comparatively insignificant. Let us suppose, for example, that at the first beat, when feebleness begins to show itself in the action of the heart, that a quarter of an ounce regurgitates as before into the auricle, and that three ounces, minus one-eighth, pass into the aorta. Now we know, that the auricle at this time contains three ounces plus a quarter of an ounce (the latter being the amount regurgitated) which are ready to be poured into the left ventricle, and when that chamber is perfectly dilated, it must contain three and a quarter ounces plus one-eighth of an ounce, which is retained. The total amount therefore is equal altogether to three and three-eighths ounces of blood; its capacity is therefore increased beyond the normal amount, by a volume, equal to three-eighths of an ounce. The heart is now stimulated to increased functional energy, expels the quarter of an ounce backward into the auricle, the three ounces into the aorta, and retains one-eighth of an ounce, and thus the three ounces of blood, which came from the right side of the heart, finds its way into the systemic circulation, and the equality of effluxes, which is absolutely essential to the maintenance of the circulation, is established. At the next and succeeding beats, indefinite in number, the same process is repeated, one-eighth of an ounce being retained at the closure of the systole, quarter of an ounce being discharged into the auricle, and three ounces being driven into the aorta, and in the process of time therefore, the functional energy develops sufficient muscular hypertrophy, to enable the heart to send on into the aorta, the exact quantity of blood, which the right ventricle sends into the lungs. The dilatation of the left ventricle is as *absolutely* essential for the production of this result, as the hypertrophy, and although looked upon by many writers as an evidence of failing of the heart, it is as important and conservative as the hypertrophy. This condition of affairs may remain, for an indefinite period of time, the abnormal opening not increasing in area, and the hygienic conditions of the patient being well maintained, and the nutrition of the muscular wall being consequently undiminished; but as such a fortunate state of things cannot continue indefinitely, the morbid opening will probably increase in area, and a larger amount of blood will regurgitate. The left ventricle becomes more and more dilated, and therefore more hypertrophied, in order to carry on the circulation. The ventricular walls may also, after a time, tend to become more feeble in their action, and therefore will expel a smaller quantity of blood, than previously into the aorta, thereby causing a larger quantity to be retained, at the close of the systole, in the ventricular chamber, and therefore eliciting a further increase of functional energy, and subsequently of further hypertrophy of the ventricle. Thus, when the mischief resulting from the diseased orifice is met by the necessary dilatation and hypertrophy of the ventricle, the circulation through the diseased heart will fairly represent that of the healthy heart. The same amount will flow simultaneously from both sides, and pulmonary congestion will be entirely prevented. We also see, that in the case where the left ventricle fails to completely empty itself, it is dilated by the amount regurgitated, plus the amount retained.

Now, in course of time, however, it is clear that increased failure of the heart's action will occur, leading to further dilatation and hypertrophy, the battle of life being fought out, step by step, until the ventricle is found to contain a large amount of residual blood, which appears for a time to be out of the pale of the circulation. The hypertrophied ventricle is able to forward into the systemic circulation whatever blood it receives from the right side of the heart. We ought not to forget, in the above consideration, the importance of the aspirating power* of the ventricle in carrying on the circulation through the cardiac chambers, in their healthy and diseased states, for it would seem, that as the left ventricle can only drive the blood to the proximity of the right auricle, where the pressure of the fluid is found to be almost negative, the blood must be drawn into the right side of the heart by the aspiration of the right ventricle, and again, as the power of the right ventricle must be mainly consumed in overcoming the resistance in the pulmonary vessels, which must be considerable in amount, we may conclude, that the blood enters the left auricle, with little or no pressure, and that it owes its capability of flowing into the left ventricle to the aspirating power of that chamber. This power of aspiration,

* Goltz and Gaule. Pflüger's *Archiv.*, xvii., 1878, p. 100.

possessed by the chambers of the heart, and by virtue of which this organ acts as a suction pump, as well as a force pump, plays evidently an important part in the circulation, and it has hitherto been too little studied in relation to diseased function. It is evident, that a diminution of it will lead to retardation in the flow of blood through its chambers.

We now come to another class of cases, in which the left auricle becomes weaker, and loses its elasticity to a certain extent, so as to permanently retain a quantity of blood, say a quarter of an ounce of the half ounce regurgitated. The auricle therefore will send into the ventricle three and a quarter ounces. At the next systole of the ventricle, two and three quarters ounces will pass into the aorta, and half an ounce will regurgitate as before. The auricle now contains three ounces from the right heart, half an ounce regurgitated, and quarter of an ounce which it permanently retains; it is therefore dilated by three and three quarters ounces. It now energises, so as to expel three and a half ounces into the ventricle, and so on. We therefore see, that although the auricle is now dilated, so as to hold three and three quarters ounces, the ventricle as before is only dilated by the amount regurgitated, namely, half an ounce. Suppose now, that the auricle again weakens and loses its elasticity, so as to retain permanently half an ounce, at the next systole of the auricle, it will expel three ounces only into the ventricle, but at the next ventricular systole, half an ounce will regurgitate as before, and two and a half ounces will pass into the aorta. The auricle will now hold three ounces from the right heart, half an ounce regurgitated and half an ounce permanently retained, which is equal altogether to four ounces. The auricle is therefore dilated to hold four ounces, but it now energises as in the case above, and drives into the ventricle the same amount as before, namely, three and a half ounces, of which, at the next systole of the ventricle, three ounces will pass into the aorta, and half an ounce will regurgitate into the auricle. We see, therefore, that although the auricle is dilated to hold four ounces, that the ventricle is still only dilated by the amount regurgitated, and so on.

Thus, if the above reasoning be correct, two main facts are shown.

1. That in this class of cases, however much the auricle is dilated to hold, the ventricle is not correspondingly dilated.

2. That by each successive weakening of the auricular wall, a certain quantity of blood is subtracted from the ventricular contents, and added to the amount permanently retained in the auricle.

The anatomical conditions, which one would expect to find present, are :—

1. Great dilatation of the left auricle and some hypertrophy of its walls, the former being permanent.

2. Some dilatation of the left ventricle with hypertrophy.

3. Hypertrophy of the right heart to supplement deficient power of the left auricle.

4. A greater tendency, than in the first class of cases, to dropsy and congestion of the lungs. It would appear, from the statements in (1) and (2) that those cases, in which there is found to be scarcely any dilatation and hypertrophy of the left ventricle, may be thus explained; for the auricle, by its gradual dilatation, prevents, as it were, the ventricle from dilating, and allows it to maintain its normal size.

Another variety of cases suggests itself. For let us suppose that the auricle, although it has been permanently dilated to hold three and three quarter ounces, gradually hypertrophies to such an extent, that it expels, not only three and a half ounces into the ventricle, but three and three quarter ounces, or, in other words, that it drives the whole of its contents into the ventricle. Then the ventricle will be dilated to hold three and three quarters ounces, instead of three and a half ounces. As it exerts the same force as before it sends at its next systole three ounces into the aorta, half an ounce into the left auricle, and retains now a quarter of an ounce. The left ventricle is therefore dilated in this case, as shown by the following considerations :—

1. By the amount regurgitated, namely, half an ounce.

2. By the amount retained, namely, quarter of an ounce.

This condition of affairs may also be brought about in another way. Let us again suppose, that the auricle retains for a time half an ounce, so that it is dilated to hold four ounces, three ounces from the right heart, half an ounce from left heart, and half an ounce retained. The auricle now gradually hypertrophies, and drives into the ventricle three and three quarters of an ounce, instead of three and a half ounces; the ventricle is therefore dilated to hold three and three quarters ounces, as in the case stated above, and as before it sends out three ounces into the aorta, half an ounce into the auricle, and retains quarter of an ounce permanently. Now suppose, again, that the auricle further weakens, and retains for a time three quarters of an ounce instead of half an ounce, it will now contain three ounces from the right heart, half an ounce regurgitated from the left ventricle, and three quarters of an ounce retained, amounting altogether to four and a quarter ounces, it is therefore now dilated by four and a quarter ounces, but let us suppose that it again gradually hypertrophies, so as to expel four ounces into the ventricle. The latter is therefore dilated to hold four ounces plus quarter of an ounce, which it permanently retains.

Two conclusions may be drawn from the above cases :—

1. However much the auricle is dilated beyond its normal size, that as soon as it hypertrophies to a sufficient extent, so as to drive almost the whole of its contents into the ventricle, leaving a small quantity behind (in the above case it is a quarter of an ounce) the ventricle is dilated to exactly the same amount as the auricle ; in fact, the auricle by its hypertrophy is able to send back into the ventricle almost the whole amount of blood, which, through the weakening of its wall, it had subtracted for a time from the ventricular contents.

The anatomical conditions, which, if the above reasoning is true, we should expect to find, are:—

1. Great dilatation and corresponding hypertrophy of the left auricle, both of which are permanent.

2. Dilatation of the left ventricle, to an almost corresponding extent, and also hypertrophy, both of which are permanent.

3. Chronic congestion of the lungs.

4. Hypertrophy of the right ventricle.

MITRAL OBSTRUCTION WITHOUT REGURGITATION.

In considering the mechanism of the circulation in this form of heart disease, we will start by estimating the forces, which serve to fill the left ventricle, bearing in mind the two fundamental laws of the mechanism of the circulation in the healthy heart, namely :—

(*a*) Synchroneity of action.

(*b*) Equality of effluxes.

Now the muscular forces, which serve to fill the left ventricle during its diastole, are :—

1. The contractile energy of the right ventricle.

2. The contractile energy of the left auricle.

3. The elastic contraction of the coats of the pulmonary artery, which materially assists also the flow of blood along this vessel and its branches. This force owes its existence to the muscular power of the right ventricle, and comes only into play as soon as the contraction of that vessel has ceased.

4. The suction power of the left ventricle.

I think, that we need not take into consideration "the interaction between blood and tissue" in the capillary system of the lungs, as we have no evidence of this power, if any, which it exerts in forwarding the circulation. There can be no doubt that the main force, which fills the left ventricle with blood, is derived from the muscular contraction of the right ventricle, the assistance from the left auricle being comparatively slight, as evidenced by the thinness of its walls, and by the very small period of time, during which the auricular contraction is called into operation. The wave, which marks the auricular contraction, starts from the immediate proximity of the vessels, which supply that chamber with blood, and passing rapidly over its walls, becomes apparently continuous with the wave of the ventricular contraction, and the succession of the two waves is so rapid, that no appreciable interval of time can be observed between the termination of the auricular and the commencement of the ventricular wave. "These two motions," says the illustrious Harvey,* "the one of the ears (auricles) and the other of the ventricles, are so done in a continued motion, as it were, keeping a certain harmony and number, that they are both done at the same time, and only one motion appears, especially in the hotter creatures, whilst they move with a sudden motion. Nor is this otherwise done, than when in engines, one wheel moving another, they all seem to move together; and in the lock of a piece, by the draw-

* "The Anatomical Exercises of Dr. William Harvey, concerning the Motion of the Heart and Blood," 1673.

ing of a spring, the flint falls, strikes the steel, fires the powder, enters the touch-hole, discharges, the ball flies out, pierces the mark, and all these motions, by reason of the swiftness of them, appear in the twinkling of an eye." Physiologists have calculated the time occupied by the auricular systole to be nearly one-eighth of the time between two consecutive pulses. If we take the pulse to beat at the rate of seventy-five per minute, the duration of the systole of the auricle would be equal to one-eighth of one-seventy-fifth of sixty seconds, or one-tenth of a second exactly. However small this interval may appear, it is evident that the auricular systole must be commenced and completed very rapidly, inasmuch as the vessels which supply the auricles are closed during the emptying of these chambers, by the contraction of the muscular fibres which surround their orifices. It would appear then, that the interior of the auricles may be shut off from communication with the pulmonic and venous systems respectively, for the small interval of time (one-tenth of a second) without injuriously affecting the circulation, and that these systems are thus protected from the reflux of blood regurgitating from the auricle, by the momentary closure of the vessels, in the manner described. Any undue prolongation of the auricular systole, by increasing the time, during which the vessels are closed, must necessarily tend to a vascular congestion more or less dangerous to life.

From these remarks, it is evident that the left auricle performs a very inferior part in filling the left ventricle, partly from the thinness of its walls, and partly from the very limited time, during which its force can be allowed to act, and as its energy only comes into play when the left ventricle is already nearly filled with blood, it is plain that its power is subsidiary to, and insignificant compared with the force exhibited by the right ventricle. There can be no doubt, however, that its muscularity aids in the complete distension of the left ventricle, as we find such *marked* hypertrophy of its walls, when the free flow of blood from its chamber, and the great vessels which supply it, is impeded as in mitral obstruction. In the normal state the auricle is never less than two-thirds full, therefore its capacity varies only by one ounce. Now as the ventricle receives three ounces during its diastole from the auricle, and at the end of its diastole, two ounces are present in the auricle, it follows, that while three ounces are passing from the auricle into the ventricle, two ounces are flowing into the auricle from the vena cava. Hence, the auricular systole takes place, when two ounces of blood have passed into the ventricle, and its action is therefore to drive an additional ounce into that chamber.

Now from our knowledge of the existence of the two fundamental laws,

1. Synchroneity of action,

2. Equality of effluxes,

we can readily predicate what must be the mechanical (physical) result, which must follow upon simple contraction of the mitral orifice *without* regurgitation. Let us suppose the area of the mitral opening to be ·25 square inches, in place of the mean normal area 1·25 square inches, the available opening being thus one-fourth of the normal. The force which would be required to drive a quantity of blood equal to the normal charge from the right ventricle and left auricle through the narrowed opening in the normal time, must be increased at least fourfold. Such a force is developed in healthy constitutions by the walls of the right ventricle augmenting in thickness, *pari passu* with the increased work required from them, aided by an augmented thickness of the left auricular walls. When the compensatory (simple) hypertrophy is established, the right ventricle sends forth its charge of three ounces into the lungs, and at the next opening of the left heart, this full charge is received into the left ventricle, which in its turn expels the three ounces, so that the ventricles are thus enabled to send out equal quantities in systole, into the pulmonic and systemic circulations respectively. There is no doubt

that this result is not obtained *per saltam;* but at
the earliest formation of the obstruction at the
mitral orifice, the right ventricle for a time,
by the instinctive power the heart possesses of
measuring the work it has to perform, increases
its muscular energy, which continues until the
growth of muscle, hypertrophy, permanently supplies the power, which converts the diseased
heart into a normally acting organ. This stage,
which consists, as has been seen, in increase of
energy of the right ventricle in the first place,
leading eventually to its hypertrophy, may be
sufficient for a long time, provided that the patient
leads a quiet life. The circulation will be evenly
maintained, the left ventricle receiving its usual
supply, but not that of full distension, as is evidenced by the comparative smallness of the pulse.
We have thus far considered the case in which
the right ventricle, aided by the left auricle, is
able to overcome the obstruction at the mitral
orifice, so as to send its charge into the left ventricle. In course of time, however, owing to the
increasing contraction at the mitral orifice, further
obstruction to the circulation is developed, so that
the right ventricle even, when aided by the left
auricle, is unequal to the task of sending its usual
supply through the obstructed orifice into the left
ventricular chamber. It is clear, that if the force
of the right ventricle and left auricle remain the

same, but the obstruction, which they have to overcome, increases, that less blood will pass through the narrowed orifice, and hence a few minims will remain behind in the left auricle. Let us suppose that sixty minims or one drachm is thus retained. At the next systole of the right ventricle, it will send its usual supply of three ounces into the left auricle, which meeting the one drachm already there, will cause the contents of that chamber to be three and one-eighths of an ounce. This chamber will therefore be dilated by this amount. But the left auricle is stimulated to increased functional energy, so that it is enabled to send into the left ventricle the exact amount which it receives from the right heart. It therefore becomes hypertrophied, as well as dilated, the former process being permanent, and the latter at first temporary. In the course of time, however, this hypertrophy will enable the left auricle to send on the whole of its contents into the left ventricle, which being naturally dilatable within certain limits, is able to receive it, and for a *brief* interval of time, therefore, more blood will enter the systemic circulation than usual. We have so far considered the case, in which although the left auricle for a time retains a certain amount of blood, it is enabled, by its hypertrophy, to overcome the obstruction, so as to send on the whole of its contents into the left ventricle.

We now pass on to consider the condition which arises, when, owing to the *increasing* obstruction at the mitral orifice, the left auricle is unable at any time to send on the *whole* of its contents ; this will lead, as shown above, to the retention of some of the charge of the right ventricle, causing further dilatation, and eventually further hypertrophy (which is now eccentric in character) of the left auricle. In progress of time, this process is frequently repeated, until the dilatation becomes so great, that the cavity of the auricle has been found capable of holding an adult fist.*

The hypertrophy of the left auricle is, however, always sufficient to send on into the left ventricle the *same* amount which it receives from the right ventricle, and thus is the law of the equality of effluxes maintained. Finally, we see how necessary and essential is the dilatation of the left auricle, in order to prevent pulmonary congestion and œdema taking place, for it is obvious that so long as this process, together with that of hypertrophy, are sufficiently developed, so long will the circulation be evenly maintained, and so long will the patient lead an unembarassed existence.

* Case recorded by Dr. Bristowe, " Pathological Society's Transactions," vol. xi., p. 66.

REGURGITANT AORTIC DISEASE.—SECTION I.

THE mechanism, by which the structurally diseased heart in such cases, is able to maintain an almost unembarassed circulation, and without the production of dropsy, is not so simple in character as the one already described. The conditions, however, which are absolutely requisite to secure this result, are the same as those, which are satisfied by the healthy heart, viz.:—

1. The synchronous action.

2. The equality of issues of the two sides of the organ.

In this, as in every other form of valvular disease, the conservative efforts of the system are powerfully directed to obtain as close an observance as possible of these fundamental laws of the circulation. The anatomical relations of the two ventricles must under all circumstances maintain the integrity of the first condition, but should the second condition fail to be secured for a certain period, the occurrence of pulmonic congestion or systemic dropsy must inevitably ensue. It is interesting to observe to what a degree insufficiency of the aortic valves may exist, and how its injurious effects may be obviated by the conservative development of the ventricle, to what an extent in fact, eccentric hypertrophy must be established in extreme cases to enable the blood to flow evenly through every part of the circulation. Let us suppose now a pure case of aortic regurgitation without obstruction of any moment at that orifice, such as for instance when rupture of a semilunar valve has been the result of excessive muscular exertion on the part of the individual. Previously to the accident, the heart had been discharging equal quantities of blood synchronously from its two ventricles, and maintaining thereby an even circulation through the lungs and body generally. The immediate effect, however, of the injury to the valve will be a reflux of blood during diastole of a portion of the charge, which had left the ventricle in systole. For the sake of example, let us suppose three ounces of blood to have been expelled by the ventricular contraction, and a quarter of an ounce to flow back into that chamber during the period of its relaxation, the systemic circulation receiving for that beat two and three quarter ounces in place of the mean normal amount, three ounces. The left ventricle, which has a certain amount of dilatability within normal limits, is compelled by the pressure of the aortic column of blood, acting through the abnormal opening of the valves, to enlarge its capacity sufficiently to contain the quarter of an ounce of reflux blood, plus the mean charge derived from the right ventricle. It contains therefore, when filled, three and a quarter ounces of blood, and its cavity is dilated consequently for the time, in excess of the mean capacity by a space equivalent to the volume of a quarter of an ounce of blood. Having now an increased quantity of blood to contract upon in systole, the energy of the left ventricle must be correspondingly increased, to enable it to expel its contents of three and a quarter ounces in the same time, in which the right ventricle sends forth its three ounces into the pulmonic circulation, for as I have frequently stated, the inter-connected muscular arrangements of the two hearts necessitate their simultaneous action. The required energy is elicited, and the ventricles are completely emptied. But as soon as the left ventricle reopens in diastole, the same process recurs. Again does it fill as before to the extent of three and a quarter ounces, and again does the renewed energy rid it of its *entire* contents, and this process, as I would term it, is repeated beat after beat, until actual development and growth of the muscular tissue follows upon persistently increased functional activity, and an amount of hypertrophy of the left ventricle is produced, sufficient to ensure the complete emptying of the chamber in systole, and thus is every charge of blood, derived from the right ventricle practically forwarded without diminution of volume, and in regular succession to the systemic circulation, and thus is the law of the equality of issues strictly satisfied. The left ventricle, when full, is of course increased in its internal capacity by the volume of a quarter of an ounce of fluid, its walls are correspondingly thickened, and the case is one of eccentric hypertrophy. The dilatation is a necessary and essential result of the reflux of the blood in diastole, for without its production, the left ventricle could not contain the normal charge sent into it from the right ventricle, plus the amount regurgitated, but by this increase in its capacity, the two chambers are enabled *practically* to send out equal volumes of blood into their respective circulations, and the injurious results from the incompetency of the valves are thus, for a time at least, completely obviated. I have used the word *practically*, meaning thereby that though three and a quarter ounces are in the case supposed, actually sent out by the left ventricle in systole, three ounces only are *virtually* added to the systemic circulation at each beat of the heart, *being the same quantity*, which the right ventricle urges at every stroke into the vessels of the lungs. The left heart therefore, *pari passu*, drains off from the pulmonary circulation the same quantity of blood per beat, which the right heart is pressing into its vessels, and thus is congestion prevented and dyspnœa undeveloped.

The eccentric hypertrophy of the left ventricle is developed to an amount, which is exactly adapted to completely empty the left ventricle, and to maintain an unembarrassed circulation.

The steps of this conservative effort upon the part of the ventricle are :—

1. Dilatation of the chamber to a degree sufficient to contain the reflux blood plus the charge sent from the right ventricle.

2. Increased functional energy of the muscular walls of the ventricle to enable it to completely expel its contents into the aorta.

3. Gradual development and growth of the ventricular wall to an amount sufficient to maintain the full power of the chamber.

The dilatation of the ventricle is therefore clearly at this stage of the affection in no way a morbid process, but is as essential to the due circulation of the blood, as the hypertrophy, which quickly follows upon it. It is evident, that the safety of the individual depends upon the due and speedy production of the hypertrophy, in fact upon the ventricle growing in muscle to meet the additional work thrown upon it. When the balance between the resistance to be overcome and the power to meet it, is established, the dangerous effects of the injured valve will be for the time counteracted, and unless the individual be exposed to undue muscular efforts, he may proceed with the ordinary duties of life "on the even tenor of his way" with but little knowledge from his feelings, of the serious lesion, which exists in his heart. How long, how many weeks or months, this quiescent condition may continue, it is impossible to conjecture, but it will continue no doubt as long as:—

1. The valvular imperfection continues unaltered.

2. The ventricular walls remain correspondingly hypertrophied, and therefore, as long as the general health is well maintained and no excessive exercise at any time attempted.

Any deviation from these conditions will lead to secondary changes, which are somewhat more complex, but of the same character as those, which lead to dilatation of the left ventricle in cases of aortic obstruction.

Should the abnormal aperture increase in area, a larger amount of blood will regurgitate into the left ventricle, and lead on the principle already explained, to the necessity of increased dilatation of the chamber, and to further augmentation in the thickness of its walls. There is no doubt, that the valvular lesion in many cases gradually increases, by reason of the centrifugal pressure of the regurgitating column of blood dilating the orifice, and thus leads consequently to a gradual and increased development of the "active dilatation " of the chamber in the manner described. But the disease progresses, I believe, usually in another manner, associated more or less with an increase in the amount of the valvular lesion. The hypertrophied left ventricle, from causes affecting the general health, becomes somewhat insufficient for its task and fails to empty itself completely at each contraction. A few minims or a drachm or more of blood remains in the cavity at the close of systole. At the next diastole, the chamber must hold the usual amount of regurgitating blood in addition to the supply sent into it from the right heart, and, although already dilated, must therefore be still further enlarged to the extent of a few minims or drachms. Increased functional energy is imperatively demanded, and perhaps the whole contents of the chamber may be sent on into the aorta, thus, for a time, preventing the development of the secondary changes. But supposing the result to be not so favourable, and that a residue of blood occurs at each systole, and that in time further hypertrophy is developed, and the already enlarged chamber is further permanently dilated to an excess of capacity equal to the volume of the minims or drachms of blood, which are so unexpelled in systole; so long as this is the case, so long is the charge of blood from the lungs sent on into the aorta, and so long is the congestion of these organs prevented and dyspnœa undeveloped. The extra dilatation and hypertrophy generated are now sufficient to the occasion, and while a certain amount of blood always remains in the ventricle (as if out of the pale of the circulation), all the blood, which the right heart sends on into the pulmonary circulation is withdrawn from it, *pari passu,* and forwarded on to the systemic circulation, and thus though the left ventricle is increased in capacity and bulk, the equality of issues from each ventricle is *practically* maintained. Thus, from time to time, failure of the left ventricle and the corresponding efforts on the part of the system to obviate the effects may continue for years, the amount of the residue gradually augmenting till the active dilatation leads to those enormous hearts, in which the left ventricle has been found to hold as much as a pint of blood.

A remarkable illustration of what I have just stated will be found in a case published by Dr. Stokes, in his work on *Diseases of the Heart and Aorta,* p. 218.

In this instance where extensive disease of the aortic valves had lasted for ten years, there had never been at any time dyspnœa, cough, or the least evidence of congestion or interrupted circulation. The post-mortem examination revealed hypertrophy and dilatation of the left ventricle to an extraordinary degree, the weight of the emptied heart being forty-four and a half ounces, and the sinuses of the aortic valves were almost filled by rugged calcareous deposits. Such developments as I have described may exist, as the above case shews, for years without the production of the slightest dropsy, such a result depending upon the fact that, however great be the mass of blood always present in the left ventricle, whatever amount enters it in diastole from the lungs, the same amount leaves it in systole to pass into the aorta. But as soon as this exact balance is destroyed, as soon as the issues from the left ventricle become persistently less than the issue from the right ventricle, so soon will pulmonary congestion commence to be developed, the right ventricle is rendered unequal to send out its *full* charge into the pulmonary vessels, diminution of flow of venous blood into the right auricle, and diminution of rapidity of current in the vessels occur, exosmosis and dropsy ensue, and all this results from the great law being unsatisfied, that equal volumes of blood must leave the right and left hearts in exactly equal and the same time.

REGURGITANT AORTIC DISEASE.—SECTION II.

In this form of cardiac mischief, the left ventricle is filled during its diastole from two sources, normally by the current, which enters through the mitral opening, and abnormally by 'the reflux stream, which the recoil of the aorta forces downwards into the chamber through the imperfectly closed orifice. The forces, which urge these simultaneously flowing currents are unequal in energy, and the openings through which the streams pass, differ from each other considerably in area. Hence, the quantities passed through each orifice, to fill the ventricle, are unequal in amount. It will be not only instructive, but interesting, to attempt to discover, in any given case, in what proportion the normal mitral and the abnormal aortic sources respectively contribute to form the ventricular charge, in fact, the quantity of blood, which occupies the chamber at the end of its diastole. The problem is by no means easy of solution, but I am indebted to the kindness of my friend, the profound mathematician, and able physiologist, Dr. Haughton, of Trinity College, Dublin, for a clear answer to the question.

I will illustrate by a rough sketch the facts of
the problem to be solved.

Let the area of M (mitral orifice) = 1·25 square
inches, and the area of the abnormal opening of
A (aortic orifice) = ·25 square inches, and sup-

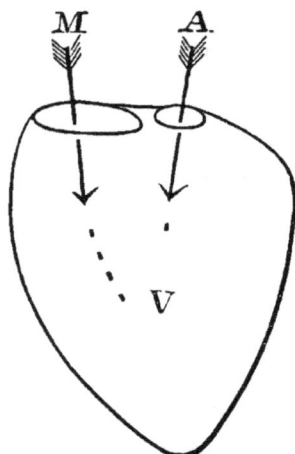

The accompanying figure, which is purely diagrammatic,
illustrates the two sources, from which the left ventricle is
supplied in aortic regurgitation. Area of normal mitral orifice
= 1·25 square inches. Area of abnormal aortic orifice = ·25
square inches. V = left ventricle. M. Mitral orifice. A.
Aortic orifice.

pose the pressure of the hæmostatic column upon
the floor of the aortic valves = weight of a column
of blood ninety inches in height, and suppose the
blood to enter through M under a pressure =
six inches column of blood, and also that the left

ventricle offers no resistance to the two incoming currents. Let the pulse = seventy-five beats per minute, and let the left ventricle contain three ounces of blood at diastole. What are the respective quantities of blood delivered through M and A, which together make up the three ounces, friction being neglected. The quantity of water discharged by a circular pipe in a given time is (including friction) determined by the following formula : —

$$Q = A\sqrt{\frac{h}{l} \times d^5}, \text{ where}$$

> A = constant.
> h = charge or head of water.
> l = length of pipe.
> d = diameter of pipe.

Let Q = quantity due to mitral orifice.
Q′ = quantity due to aortic orifice.

$$\frac{Q}{Q'} = \sqrt{\frac{h}{h'} \times \frac{l'}{l} \times \frac{d^5}{d'^5}}$$

We may neglect the difference between l and l' in the present problem.

Hence $\dfrac{Q}{Q'} = \sqrt{\dfrac{h}{h'} \times \dfrac{d^5}{d'^5}}$

h = 6 inches and h' = 90 inches.

Now $\dfrac{d^2}{d'^2} = \dfrac{1\cdot 25}{\cdot 25} = 5$, as the area varies as the square of the diameter.

E

$$\therefore \frac{d^6}{d'^6} = 55\cdot902, \text{ and therefore}$$

$$\frac{Q}{Q'} = \sqrt{\frac{6 \times 55\cdot902}{90}} = \sqrt{3\cdot727} = 1\cdot930.$$

Hence the quantity of blood, which enters the ventricle through the mitral orifice, is 1·930 times as much as the quantity regurgitated through the aortic opening. The three ounces will therefore be divided in the following proportion :—

$$Q + Q' : Q' : : 3 : \text{quantity regurgitated.}$$

$$\therefore \text{ Aortic portion} = \frac{3\,Q'}{Q + Q'} = \frac{3}{\dfrac{Q}{Q'} + 1} = \frac{3}{2\cdot930}$$

$$\therefore \text{ Aortic supply} = 1\cdot024$$
$$\text{Mitral supply} = 1\cdot976$$
$$\overline{3\cdot000}\text{ ounces.}$$

In the above case, the openings, normal and abnormal, have been supposed to be circular in form. The *mitral* orifice *in action* is, I think, always circular, but the diseased aortic opening is rarely of this form, being usually irregular in outline. The problem, which I have proposed, and which Professor Haughton has so ably solved, does not aim of course at being of any especial practical value, but simply enables us to form a clear estimate of how the ventricle would be filled, should certain conditions, as to pressure and area, be given. It is seen, that with the data of this

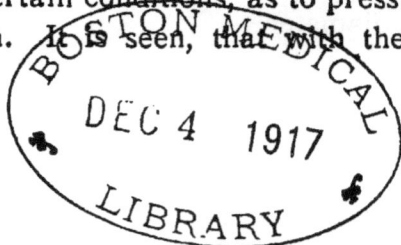

particular case, the aortic column supplies rather more than one-third, and the direct mitral rather less than two-thirds of the whole ventricular contents, which are to be discharged into the aorta at the next systole of that chamber. The pressure of six inches at the mitral orifice is simply hypothetical, but that this pressure is small, is clear from the experiments, which have been made upon the pressure of the blood entering the right ventricle in animals, a pressure which cannot be very much augmented by the contraction of the auricular chambers, which are so thin in their muscular walls, and which act only during a very small portion of the time, from pulse to pulse, being, as calculated above, one-tenth of a second.

CONTRACTION OF THE AORTIC ORIFICE WITHOUT REGURGITATION.

THE mechanism of the circulation in this case is very simple. In the healthy human heart, the velocities of the streams, which issue simultaneously from the right and left ventricles, are nearly in the ratio of three to four, inasmuch as the two chambers have to send out equal volumes of blood synchronously through the pulmonic and aortic apertures, whose mean areas are nearly one square inch, and three-fourths of a square inch respectively— neglecting friction. A contracted aortic orifice

E 2

usually presents a very irregular form, but supposing, for the sake of example, its effective opening to be circular in shape, and its diameter one
half of that of the normal outlet, the force of the
ventricle must be increased at least fourfold to
enable the contents of the ventricle to be discharged in the usual period of systole; further,
the greatly increased friction due to the roughened
orifice must also increase the necessary force of the
ventricle.   Should this power be developed and
maintained permanently by an hypertrophy of the
muscular walls of the ventricle, the two fundamental
conditions of synchronism of action, and equality
of issues from the two ventricles, will be satisfied,
and the circulation enabled to proceed with but
little embarassment to the individual.   Both ventricles will be synchronously completely emptied
of their contents, and all pulmonic congestion
avoided, unless by undue muscular exertion on
the part of the individual, an excess of blood be
poured into the left chamber.   In such a case, the
ventricle will become for a time abnormally distended, and unable to free itself entirely, at each
systole, of its contained blood, until cessation from
bodily exertion enables the right and left hearts
to regain the equilibrium, which had been dis-
turbed.   It is also evident, that a passing dilatation
of the ventricle will occur.   I have supposed the
left ventricle to have exactly adapted itself, by the

simple hypertrophy of its walls, to the obstruction,
which it has to overcome at the aortic orifice, and
that no permanent distension has taken place.
The contraction has so far had no marked effect
on the cardiac circulation, beyond rendering the
period of systole somewhat prolonged, in com-
parison with the period of diastole. This is
clearly proved by the sphygmographic tracing of
the radial pulse, where the unusually oblique and
curved line of the ascent indicates the difficulty,
which the blood had experienced in forcing its
way through the abnormally contracted opening.
Let us suppose, however, that the hypertrophied
wall of the ventricle becomes diminished in energy,
from functional or organic causes, and unequal to
the task, which it has hitherto been able to accom-
plish, viz., the complete emptying of the chamber.
A small quantity of blood, it may be at the outset
a few minims only, or a drachm or two, remain in
the ventricle at the close of its systole. As the
right side of the heart, aided by the left auricle,
continues to propel the usual quantity of blood into
the left ventricle, that chamber is compelled to
yield to the pressure of the current through the
mitral orifice, and becomes consequently sufficiently
dilated to contain the normal charge, from the
auricle, plus the quantity which had been left in
it from the previously incomplete systole. There
are now, we may say, three and a quarter in place

of three ounces of blood in the left ventricle, and the capacity of the cavity is increased by a quarter of an ounce of fluid. The ventricle, which had become already unequal to its normal task, has now to act with increased force upon a larger volume of blood, and must become further hyper-trophied, in order that the three ounces, which it had received from the auricle, may be forwarded, without hindrance, through the aortic orifice. Eccentric hypertrophy now commences, and should a sufficiently compensatory thickening of the walls be established, three ounces of blood will be syn-chronously discharged from the two sides of the heart, and a quarter of an ounce will remain in the left ventricle, at the close of each systole, while the amount of permanent dilatation of the chamber will equal the volume of a quarter of an ounce of fluid, and thus will the laws of synchronism of action, and equality of issues, be satisfied. This amount (quarter of an ounce) is of course merely hypothetical, and cited as a means of illustrating the manner in which the struggle between the obstruction and the compensatory hypertrophy can be so balanced, that the circulation (pulmonic and systemic) may, for the time, be efficiently maintained, although a portion of the blood (as if out of the pale of the circulation) is constantly retained in the ventricle, behind the contracted aortic orifice. It is evidently impossible to speculate, even upon the length of the period, during which this armistice between the two antagonising powers can be prolonged, and further dilatation of the ventricle prevented. A heart moderately dilated, as the result of this affection, may remain unaltered in its internal capacity for months, and even years, provided that no excessive quantity of blood be too frequently forced into its right chamber by undue muscular exercise, and that the nutritive and contractile energy of the muscular walls be maintained, and provided also, that no increase takes place in the amount of the obstruction at the diseased outlet. If, however, one of these conditions is imperfectly fulfilled, the left ventricle (which, as already stated, failed to empty itself completely) will again discharge a still smaller proportion of its contents into the aorta. Let us suppose a quarter of an ounce to be retained in the ventricle, in addition to the quantity, which had already been persistently left in that chamber after its systole. Half an ounce of blood will therefore remain in the ventricle at the close of its contraction, producing a corresponding permanent increase in its capacity, and further dilatation of its cavity can only be prevented by an increased hypertrophy of its walls. At this point, therefore, when a balance has been once more established between the obstruction and the hypertrophy, the quantity (say three ounces) of blood, which has entered the left ventricle from the auricle, is enabled to be completely discharged through the narrowed outlet of the aorta, and in this way the two sides of the heart expel synchronously equal quantities of blood, and half an ounce of blood remains in the left ventricle, producing a corresponding amount of permanent dilatation of that cavity. The mechanism by which the eccentric hypertrophy is produced, is very gradual in its action, and slow in effecting its result, and does not lead to the enlargement of the heart *per saltam*, as might be inferred from the account just given. I merely described these stages as periods in the development of " actual dilatation," where the obstruction and the counteracting hypertrophy actually balance each other, a sa mode of illustrating generally the nature of the process in operation. The force, which dilates the left ventricle, is entirely derived from the contraction of the right ventricle and left auricle (as I consider the "pulmonary interaction force " to be of but little avail as a power of the circulation) aided by the suction power of the ventricle itself, due to the elasticity of its fibres, by means of which the walls spring apart in diastole. Now even under the most favourable circumstances, this force is of comparatively feeble energy. We must remember, however, that it acts upon the internal surface of the ventricle, during the period of its muscular relaxation,

and will be of course more effectual in producing
enlargement, when the walls of that chamber are
deficient in power, from functional or organic
causes. There can be no doubt then, that the
production of the eccentric hypertrophy of the left
ventricle is the result of a series of stages, occur-
ring at very unequal intervals of time, and that
the parts of each stage are as follows:—

1. Retention of a small quantity of blood at the
close of the systole, from inability of the ventricle
to overcome completely the resistance at the contracted aortic orifice.

2. Over-distension of the ventricle, from the
presence within it of the above residue in addition
to the usual volume of blood received by it in
diastole from the right side of the heart.

3. Increased, but insufficient effort on the part
of the ventricle to expel its entire contents, a quantity of blood remaining at the end of each systole,
equal to or less than the residue from the previous
systole. In other words, the increased energy of
the ventricle being sufficient to prevent any further
over-distension of the chamber, and to carry forward into the aorta the entire charge received by
it in diastole from the auricle.

4. Persistence of the ventricle in this increased
functional energy, until sufficient hypertrophy of
the ventricular wall shall have been developed to
maintain permanently the increased demand on its
contracting power.

5. And as a consequence of 4, the force of the ventricle will then exactly counteract the resistance at the aortic orifice, and the circulation through the heart be thoroughly established, although a certain amount of dilatation remains constant.

The production of the hypertrophy is, I need scarcely say, an example of the familiar physiological law, that the increased use of a muscle or set of muscles leads to their increased volume and power, provided that their tissue is healthy and sound. An increase, from time to time, in the obstruction at the aortic orifice will of course lead to a corresponding repetition of the process just described, and produce further dilatation and compensatory hypertrophy, and thus will the constant struggle between the resistance at the outlet and the contractile energy of the ventricle give rise to marked eccentric hypertrophy of the left side of the heart. I have dwelt at some length upon this point, as I find that Niemeyer in his excellent *Text-book of Practical Medicine* has failed to see the mechanism of the circulation in this class of cases. He says: —" In simple stricture of the aortic valve, the left ventricle has no increase of pressure to support during diastole, and hence does not *become dilated ;* it has, however, to propel the blood through a contracted orifice, and becomes hypertrophied on account of the greater amount of effort thus required from it. In contradistinction, then, to what

we meet with in insufficiency of the valves, we find
a simple hypertrophy instead of eccentric hypertrophy of the left ventricle, when the aortic valves
are contracted."◊   Cases of *simple* aortic obstruction are rarely met with, as the affection which
causes it usually produces also regurgitation
through the orifice.   But the following instance nar-
rated in the "Pathological Society's Transactions"
of the year 1870, is one in point.   The patient was
under the care of Dr. Goodfellow in the Middlesex
Hospital, and died December 29, 1869.   " I have
here a specimen" says Dr. John Murray, "shewing
the extent to which aortic valvular narrowing
may reach, and yet be compatible with life.   The
orifice of the valve is reduced to a mere chink, and
this appeared even still smaller in the recent state ;
in fact it was only on stretching the valve, that any
opening whatever could be observed looking from
above downward.   The free edges of the valves
are in a state of ulceration, and on them is some
fibrinous deposit, which from its firmness does not
give one the idea of being of very recent origin ;
but without this deposit, the aortic orifice from the
rigidity of the valves is greatly narrowed.   There
was marked hypertrophy of the heart, especially
of the walls of the left ventricle, *this cavity was also
much dilated.*"   A similar case to this is related by
Dr. Robert King in the " Pathological Society's

* English Edition, page 346.

Transactions," 1873, vol. xxiv. He states, " on passing the finger into the aorta, a nodulated mass of calcareous hardness could be felt in the situation of the aortic valves, but no opening into the ventricle could be detected by the touch. On slitting open the aorta down to the level of the valves, the aortic opening was found to be almost completely occluded by the hard mass above mentioned, which was merely perforated near its centre by an opening, which would barely admit a number six catheter, and which was so effectually overgrown by vegetations, that water would not pass either from the aorta into the ventricle, or from the ventricle into the aorta without considerable pressure. The left ventricle itself was *much dilated* and its walls greatly hypertrophied. The mitral and and tricuspid valves were slightly thickened and opaque, but apparently competent." These cases clearly exemplify the error of Niemeyer's statement. Although the organic lesion had evidently lasted for years in both cases, the œdematous swelling of the legs in the first one was only slight; nor was it apparently well marked in the second, and in both, the present illness had only commenced about two months before the fatal termination. Again in the first case, an attack of inflammation of the upper part of the left lung, shewn post-mortem, by the presence of red hepatisation, interfered no doubt with the action of the

right ventricle, and impeded the flow of blood from
that chamber to the left side of the heart. Hence
followed the rapid (136 beats per minute) irregular
and unequal pulse, quick respirations and imperfect
supply of blood to the nervous centres, and death.
The production of eccentric hypertrophy in the
manner, which I have described, is, I believe, the
necessary consequence of prolonged aortic obstruction, but the amount to which it may become
developed, even in the most serious cases, con-
trasts strongly with the enormous dilatation and
thickening of the walls of the left ventricle usually
found in cases of aortic regurgitation. The difference in the result of the two forms of aortic
disease depends mainly upon the great disparity in
the pressure power of the mitral direct and aortic
reflux currents, which tend to dilate the left ventricle during the period of its relaxation.

ON THE FOUR CHIEF ORIFICES OF THE HEART.

In the foregoing account, I have taken for granted
that the four openings are circular during function, and I may for a few moments dwell upon
the proofs. All writers agree in describing the
aortic and pulmonic orifices to be circular in
shape, we have, therefore, only to shew, that the
mitral and tricuspid are circular, although usually

described as elliptic in form. At the outset I may state that the canals of the body and all the outlets through which fluid has to pass, are, with a few exceptions, circular in form, and as the largest quantity of fluid can pass with the least amount of friction through a circular opening, it is clear therefore why this form is adopted. If a square and an ellipse and a circle each contain one square inch, their respective perimeters are 4, 3·85, 3·54 inches. An elliptic mitral orifice would have a perimeter one third larger than the perimeter of a mitral opening containing exactly the same area, but of a circular form.

The elliptic would therefore expose a much larger frictional surface, and would therefore be less favourable to the transit of blood than an opening of exactly circular form enclosing the same area. If the aortic and pulmonic openings are circular, as admitted by all observers, why should the mitral and tricuspid openings be less favourably formed for the passage of the currents, which traverse them. Post-mortem, the mitral and tricuspid openings present any shape but the purely circular as the heart is lax and empty, but when during life the currents of blood are flowing, the centrifugal pressure exerted by the inflowing blood must necessarily tend to throw the openings into a circular shape. By filling the heart with fluid plaster of Paris, the circularity of the openings is easily shewn. In the Bison Europæus the walls of the tricuspid and mitral orifices are so thick, that the circular shape of the opening is readily observed post-mortem. Lastly, Dr. Peacock employed spherical balls for the measurements of the openings, and found them well adapted for that shape. It is equally an interesting fact, upon which no stress has been laid that each of the four openings is unaltered and unalterable *in its area* during systole or diastole, as is shewn by the following reasons :—

(*a*) The rings, which surround the openings are formed of white, fibrous, inelastic tissue.

(*b*) The muscular fibres around these openings are not arranged in the form of a sphincter, their ultimate insertion being at right angles to the plane of the orifices. The dissections of Dr. Pettigrew* amply prove this statement.

(*c*) Mr. Savory† has shewn, that no arrangements of circular fibres exist, capable of altering in any way the areas of the tricuspid and mitral orifices; therefore, since neither of the openings are capable of being affected by muscular action, their areas must always be constant. The necessity for this condition is shewn by the following reasons :—

1. The incoming currents have a minimum

* J. Pettigrew, "Philosophical Transactions," 1864.
† W. Savory, "Royal Society's Transactions," 1851.

amount of friction to overcome, for if the openings are closed by sphincter action, a considerable amount of auricular force would have been required to ensure the passage of the blood stream through them.

2. Could the openings be closed by the muscular contractions, the arrangement of the tricuspid and mitral curtains would not only be completely unnecessary, but prove an obstacle to the circulation. In fact, where sphincters exist, no valves are present, as shewn in the case of the pylorus, anus, orifice of the bladder in man and the right auriculo-ventricular orifice in birds.

SELECTED LIST

OF

NEW AND RECENT WORKS

PUBLISHED BY

H. K. LEWIS,

136 GOWER STREET, LONDON, W.C.

(ESTABLISHED 1844)

*_*_ *For full list of works in Medicine and Surgery published by H. K. Lewis see complete Catalogue sent post free on application.*

SIR WILLIAM AITKEN, KNT., M.D., F.R.S.
Professor of Pathology in the Army Medical School.

ON THE ANIMAL ALKALOIDS, THE PTOMAINES, LEUCOMAINES, AND EXTRACTIVES IN THEIR PATHOLOGICAL RELATIONS. Second edition, crown 8vo, 3s. 6d. [*Now ready.*

H. RADCLIFFE CROCKER, M.D. LOND., B.S., F.R.C.P.
Physician, Skin Department, University College Hospital.

DISEASES OF THE SKIN: THEIR DESCRIPTION, PATHOLOGY, DIAGNOSIS, AND TREATMENT. With Illustrations, 8vo, 21s.

ANGEL MONEY, M.D., F.R.C.P.
Assistant Physician to University College Hospital, and to the Hospital for Sick Children, Great Ormond Street.

THE STUDENT'S TEXTBOOK OF THE PRACTICE OF MEDICINE. Fcap. 8vo, 6s. 6d.

LOUIS C. PARKES, M.D., D.P.H. LOND. UNIV.
Assistant Professor of Hygiene, University College, London; Fellow and Member of the Board of Examiners of the Sanitary Institute; Assistant Examiner in Hygiene, Science and Art Department South Kensington.

HYGIENE AND PUBLIC HEALTH. Second Edition. With Illustrations, crown 8vo, 9s. [*Now ready.* *Lewis's Practical Series.*]

HENRY R. SWANZY, A.M., M.B., F.R.C.S.I.
Examiner in Ophthalmic Surgery at the Royal College of Surgeons, Ireland; Surgeon to the National Eye and Ear Infirmary, Dublin, etc.

A HANDBOOK OF DISEASES OF THE EYE AND THEIR TREATMENT. Third Edition, Illustrated with Wood Engravings, Colour Tests, etc., large post 8vo

4000—5'90.

E. CRESSWELL BABER, M.B. LOND.
Surgeon to the Brighton and Sussex Throat and Ear Dispensary.

A GUIDE TO THE EXAMINATION OF THE NOSE WITH REMARKS ON THE DIAGNOSIS OF DIS- EASES OF THE NASAL CAVITIES.   With Illustrations, small 8vo, 5s. 6d.

JAMES B. BALL, M.D. (LOND.), M.R.C.P.
Physician to the Department for Diseases of the Throat and Nose, and Senior Assistant Physician, West London Hospital.

A HANDBOOK OF DISEASES OF THE NOSE AND NASO-PHARYNX.   With Illustrations, large post 8vo, 6s.   [*Nearly ready.*

G. GRANVILLE BANTOCK, M.D., F.R.C.S. EDIN.
Surgeon to the Samaritan Free Hospital for Women and Children.

I.

RUPTURE OF THE FEMALE PERINEUM.   Second Edition, with Illustrations, 8vo, 3s. 6d.

II.

ON THE USE AND ABUSE OF PESSARIES. With Illustrations, Second Edition, 8vo, 5s.

ASHLEY W. BARRETT, M.B. LOND., M.R.C.S., L.D.S.E.
Dental Surgeon to, and Lecturer on Dental Surgery in the Medical School of, the London Hospital.

DENTAL SURGERY FOR MEDICAL PRACTI- TIONERS AND STUDENTS OF MEDICINE.   Second edition, with Illustrations, cr. 8vo, 3s. 6d.   [*Now ready.* *Lewis's Practical Series.*]

FANCOURT BARNES, M.D., M.R.C.P.
Physician to the Chelsea Hospital; Obstetric Physician to the Great Northern Hospital, &c.

A GERMAN-ENGLISH DICTIONARY OF WORDS AND TERMS USED IN MEDICINE AND ITS COGNATE SCIENCES.   Square 12mo, Roxburgh binding, 9s.

H. CHARLTON BASTIAN, M.A., M.D., F.R.S.
Examiner in Medicine at the Royal College of Physicians; Physician to University College Hospital, etc.

PARALYSES: CEREBRAL, BULBAR, AND SPI- NAL.   A Manual of Diagnosis for Students and Practi- tioners.   With numerous Illustrations, 8vo, 12s. 6d.

E. H. BENNETT, M.D., F.R.C.S.I.
Professor of Surgery, University of Dublin.
AND
D. J. CUNNINGHAM, M.D., F.R.C.S.I.
Professor of Anatomy and Chirurgery, University of Dublin.

THE SECTIONAL ANATOMY OF CONGENITAL CŒCAL HERNIA. With coloured plates, sm. folio, 5s. 6d.

HORATIO R. BIGELOW, M.D.
Permanent Member of the American Medical Association; Fellow of the British Gynæcological Society, etc.

GYNÆCOLOGICAL ELECTRO-THERAPEUTICS. With an introduction by Dr. Georges Apostoli. With Illustrations, demy 8vo, 8s. 6d.

DRS. BOURNEVILLE AND BRICON.

MANUAL OF HYPODERMIC MEDICATION. Translated from the Second Edition, and Edited, with Therapeutic Index of Diseases, by Andrew S. Currie, M.D. Edin., etc. Crown 8vo, 6s.

STEPHEN S. BURT, M.D.
Professor of Clinical Medicine and Physical Diagnosis in the New York Postgraduate Medical School and Hospital.

EXPLORATION OF THE CHEST IN HEALTH AND DISEASE. Post 8vo, 6s.   [*Just published.*

DUDLEY W. BUXTON, M.D., B.S., M.R.C.P.
Administrator of Anæsthetics to University College Hospital and to the Hospital for Women, etc.

ANÆSTHETICS THEIR USES AND ADMINISTRATION. Second Edit., with Illustrations, crown 8vo. *Lewis's Practical Series.*]

HARRY CAMPBELL, M.D., B.S. LOND.
Member of the Royal College of Physicians; Assistant Physician and Pathologist to the North-West London Hospital.

I.

THE CAUSATION OF DISEASE. An exposition of the ultimate factors which induce it. Demy 8vo, 12s. 6d.

II.

FLUSHING AND MORBID BLUSHING, THEIR PATHOLOGY AND TREATMENT. With plates and wood engravings, royal 8vo, 10s. 6d.   [*Now ready.*

ALFRED H. CARTER, M.D. LOND.
Member of the Royal College of Physicians; Physician to the Queen's Hospital, Birmingham, &c.

ELEMENTS OF PRACTICAL MEDICINE.   Fifth Edition, crown 8vo, 9s.

P. CAZEAUX.
Adjunct Professor in the Faculty of Medicine of Paris, &c.

AND

S. TARNIER.
Professor of Obstetrics in the Faculty of Medicine of Paris.

OBSTETRICS: THE THEORY AND PRACTICE; including the Diseases of Pregnancy and Parturition, Obstetrical Operations, &c. Seventh Edition, with plates, and wood-engravings, royal 8vo, 35s.

FRANCIS HENRY CHAMPNEYS, M.A., M.B. OXON., F.R.C.P.
Obstetric Physician and Lecturer on Obstetric Medicine at St. George's Hospital; Examiner in Obstetric Medicine in the University of London, etc.

EXPERIMENTAL RESEARCHES IN ARTIFICIAL RESPIRATION IN STILLBORN CHILDREN, AND ALLIED SUBJECTS. Crown 8vo, 3s. 6d.

W. BRUCE CLARKE, M.A., M.B. OXON., F.R.C.S.
Assistant Surgeon to, and Senior Demonstrator of Anatomy and Operative Surgery at St. Bartholomew's Hospital, &c.

THE DIAGNOSIS AND TREATMENT OF DISEASES OF THE KIDNEY AMENABLE TO DIRECT SURGICAL INTERFERENCE. Demy 8vo, with Illustrations, 7s. 6d.

WALTER S. COLMAN, M.B. LOND.
Formerly Assistant to the Professor of Pathology in the University of Edinburgh.

SECTION CUTTING AND STAINING: A Practical Guide to the Preparation of Normal and Morbid Histological Specimens. Illustrations, crown 8vo, 3s.

ALFRED COOPER, F.R.C.S.
Surgeon to the St. Mark's Hospital for Fistula and other Diseases of the Rectum.

A PRACTICAL TREATISE ON DISEASES OF THE RECTUM. Crown 8vo, 4s.

W. H. CORFIELD, M.A., M.D. OXON.
Professor of Hygiene and Public Health in University College, London.

DWELLING HOUSES: their Sanitary Construction and
Arrangements.   Second Edition, with Illustrations, crown
8vo, 3s. 6d.

EDWARD COTTERELL, M.R.C.S. ENG., L.R.C.P. LOND.

ON SOME COMMON INJURIES TO LIMBS: their
Treatment and After-Treatment including Bone-Setting (so-
called).   Imp. 16mo, with Illustrations, 3s. 6d.

CHARLES CREIGHTON, M.D.

I.
ILLUSTRATIONS OF UNCONSCIOUS MEMORY
IN DISEASE, including a Theory of Alteratives.   Post
8vo, 6s.

II.
CONTRIBUTIONS TO THE PHYSIOLOGY AND
PATHOLOGY OF THE BREAST AND LYMPHA-
TIC GLANDS.   Second Edition, illustrated, 8vo, 9s.

III.
BOVINE TUBERCULOSIS IN MAN: An Account of
the Pathology of Suspected Cases.   With Chromo-lithographs and other Illustrations, 8vo, 8s. 6d.

EDGAR M. CROOKSHANK, M.B. LOND., F.R.M.S.
Professor of Bacteriology, King's College, London.

HISTORY AND PATHOLOGY OF VACCINATION.
2 vols., royal 8vo, coloured plates, 36s.

RIDLEY DALE, M.D., L.R.C.P. EDIN., M.R.C.S. ENG.

EPITOME OF SURGERY.   Large 8vo, 10s. 6d.

HERBERT DAVIES, M.D., F.R.C.P.
Late Consulting Physician to the London Hospital, and formerly Fellow of
Queen's College, Cambridge.

THE MECHANISM OF THE CIRCULATION OF
THE BLOOD THROUGH ORGANICALLY DIS-
EASED HEARTS.   Edited by ARTHUR TEMPLER DAVIES, B.A.
M.B. Cantab., M.R.C.P.   Crown 8vo, 3s. 6d.

HENRY DAVIS, M.R.C.S. ENG.
Teacher and Administrator of Anæsthetics to St. Mary's and the National Dental Hospitals.

G UIDE TO THE ADMINISTRATION OF ANÆS- THETICS. Fcap. 8vo, 2s.

AUSTIN FLINT, M.D., LL.D.
Professor of Physiology and Physiological Anatomy at the Bellevue Hospital Medical College, New York, &c., &c.

A TEXT-BOOK OF HUMAN PHYSIOLOGY. Fourth edition, with 316 illustrations, royal 8vo, 25s.
[Just published.

J. MILNER FOTHERGILL, M.D.
I.

A MANUAL OF DIETETICS. Large 8vo, 10s. 6d.

II.

T HE HEART AND ITS DISEASES, WITH THEIR TREATMENT; INCLUDING THE GOUTY HEART. Second Edition, entirely re-written, copiously illustrated with woodcuts and lithographic plates, 8vo, 16s.

III.

I NDIGESTION, BILIOUSNESS, AND GOUT IN ITS PROTEAN ASPECTS.
PART I.—INDIGESTION AND BILIOUSNESS. Second Edition, post 8vo, 7s. 6d.
PART II.—GOUT IN ITS PROTEAN ASPECTS. Post 8vo, 7s. 6d.

IV.

T HE TOWN DWELLER: HIS NEEDS AND HIS WANTS. Post 8vo, 3s. 6d.   *[Now ready.*

FORTESCUE FOX, M.D. LOND.
Fellow of the Medical Society of London.

S TRATHPEFFER SPA, ITS CLIMATE AND WATERS, with observations, historical, medical, and general, descriptive of the vicinity. Illustrated, cr. 8vo, 2s. 6d. *nett.*
[Just published.

ALFRED W. GERRARD, F.C.S.
Examiner to the Pharmaceutical Society; Teacher of Pharmacy and Demonstrator of Materia Medica to University College Hospital, etc.

E LEMENTS OF MATERIA MEDICA AND PHAR- MACY. Crown 8vo, 8s. 6d.

JOHN GORHAM, M.R.C.S.
Fellow of the Physical Society of Guy's Hospital, etc.

TOOTH EXTRACTION : A manual of the proper mode of extracting teeth, with a Table exhibiting the names of all the teeth, the instruments required for extraction and the most approved methods of using them.   Third edition, fcap 8vo, 1s. 6d.

[*Just published.*

GEORGE M. GOULD, A.B., M.D.
Ophthalmic Surgeon to the Philadelphia Hospital, etc.

A NEW MEDICAL DICTIONARY. A compact concise Vocabulary, convenient for reference, based on recent medical literature.   Small 8vo, over 500 pp., 16s.

[*Now ready.*

J. B. GRESSWELL, M.R.C.V.S.
Provincial Veterinary Surgeon to the Royal Agricultural Society.

VETERINARY PHARMACOLOGY AND THERA-PEUTICS.   Fcap. 8vo, 5s.

BERKELEY HILL, M.B. LOND., F.R.C.S.
Professor of Clinical Surgery in University College; Surgeon to University College Hospital, and to the Lock Hospital.

I.

THE ESSENTIALS OF BANDAGING. For Managing Fractures and Dislocations ; for administering Ether and Chloroform ; and for using other Surgical Apparatus.   Sixth Edition, with Illustrations, fcap. 8vo, 5s.

II.

CHRONIC URETHRITIS, AND OTHER AFFEC-TIONS OF THE GENITO-URINARY ORGANS. Royal 8vo, coloured plates, 3s. 6d.

BERKELEY HILL, M.B. LOND., F.R.C.S.
Professor of Clinical Surgery in University College.
AND
ARTHUR COOPER, L.R.C.P., M.R.C.S.
Surgeon to the Westminster General Dispensary, &c.

I.

SYPHILIS AND LOCAL CONTAGIOUS DISOR-DERS. Second Edition, entirely re-written, royal 8vo, 18s.

II.

THE STUDENT'S MANUAL OF VENEREAL DIS-EASES. Being a Concise Description of those Affections and of their Treatment.   Fourth Edition, post 8vo, 2s. 6d.

NORMAN KERR, M.D., F.L.S.
President of the Society for the Study of Inebriety ; Consulting Physician,
Dalrymple Home for Inebriates, etc.

INEBRIETY: ITS ETIOLOGY, PATHOLOGY,
TREATMENT, AND JURISPRUDENCE. Second Edition, crown 8vo, 12s. 6d. [*Now ready*

J. WICKHAM LEGG, F.R.C.P.
Assistant Physician to Saint Bartholomew's Hospital, and Lecturer on
Pathological Anatomy in the Medical School.

A GUIDE TO THE EXAMINATION OF THE
URINE ; intended chiefly for Clinical Clerks and Students. Sixth Edition, revised and enlarged, with additional Illustrations, fcap. 8vo, 2s. 6d.

LEWIS'S POCKET CASE BOOK FOR PRACTITIONERS AND STUDENTS. Designed by A. T.
BRAND, M.D. Roan, with pencil, 3s. 6d. *nett.*

LEWIS'S POCKET MEDICAL VOCABULARY.
Over 200 pp., 32mo, limp roan, 3s. 6d.

T. R. LEWIS, M.B., F.R.S. ELECT, ETC.
Late Fellow of the Calcutta University; Surgeon-Major Army Medical Staff.

PHYSIOLOGICAL AND PATHOLOGICAL RE-
SEARCHES. Arranged and edited by SIR WM. AITKEN,
M.D., F.R.S., G. E. DOBSON, M.B., F.R.S., and A. E. BROWN,
B.Sc. Crown 4to, portrait, 5 maps, 43 plates including 15 chromolithographs, and 67 wood engravings, 30s. *nett.*

WILLIAM THOMPSON LUSK, A.M., M.D.
Professor of Obstetrics and Diseases of Women in the Bellevue Hospital
Medical College, &c.

THE SCIENCE AND ART OF MIDWIFERY. Third
Edition, revised and enlarged, with numerous Illustrations,
8vo, 18s.

LEWIS'S PRACTICAL SERIES.

These volumes are written by well-known Hospital Physicians and Surgeons recognised as authorities in the subjects of which they treat. They are of a thoroughly Practical nature, and calculated to meet the requirements of the general Practitioner and Student and to present the most recent information in a compact and readable form; the volumes are handsomely got up and issued at low prices, varying with the size of the works.

HYGIENE AND PUBLIC HEALTH.
By LOUIS C. PARKES, M.D., D.P.H. Lond. Univ., Assistant Professor of Hygiene, University College, London; Fellow, and Member of the Board of Examiners of the Sanitary Institute; Assistant Examiner in Hygiene, Science and Art Department South Kensington. Second Edition, with Illustrations, cr. 8vo., 9s. [Now ready.

MANUAL OF OPHTHALMIC PRACTICE.
By C. HIGGENS, F.R.C.S , Ophthalmic Surgeon to Guy's Hospital; Lecturer on Ophthalmology at Guy's Hospital Medical School. Illustrations, crown 8vo, 6s.

A PRACTICAL TEXTBOOK OF THE DISEASES OF
WOMEN. By ARTHUR H. N. LEWERS, M.D. Lond., M.R.C.P. Lond., Assistant Obstetric Physician to the London Hospital; Examiner in Midwifery and Diseases of Women to the Society of Apothecaries of London; etc. Second Edition, with Illustrations, crown 8vo, 9s.

ANÆSTHETICS THEIR USES AND ADMINISTRATION.
By DUDLEY W. BUXTON, M.D., B.S., M.R.C P., Administrator of Anæsthetics to University College Hospital and to the Hospital for Women, Soho Square. Second Edition, crown 8vo. [Nearly ready.

TREATMENT OF DISEASE IN CHILDREN: EMBODYING
THE OUTLINES OF DIAGNOSIS AND THE CHIEF PATHOLOGICAL DIFFERENCES BETWEEN CHILDREN AND ADULTS. By ANGEL MONEY, M.D., F.R.C.P., Assistant Physician to the Hospital for Children, Great Ormond Street, and to University College Hospital. Second Edition, crown 8vo, 10s. 6d.

ON FEVERS: THEIR HISTORY, ETIOLOGY, DIAGNOSIS,
PROGNOSIS, AND TREATMENT. By ALEXANDER COLLIE, M.D. (Aberdeen), Medical Superintendent of the Eastern Hospitals. Coloured plates, cr. 8vo, 8s. 6d.

HANDBOOK OF DISEASES OF THE EAR FOR THE
USE OF STUDENTS AND PRACTITIONERS. By URBAN PRITCHARD, M.D. (Edin.), F.R.C.S. (Eng.), Professor of Aural Surgery at King's College, London; Aural Surgeon to King's College Hospital. With Illustrations, crown 8vo, 4s. 6d

A PRACTICAL TREATISE ON DISEASES OF THE KID-
NEYS AND URINARY DERANGEMENTS. By C. H. RALFE, M.A., M.D. Cantab., F.R.C.P. Lond., Assistant Physician to the London Hospital. With Illustrations, crown 8vo, 10s. 6d.

DENTAL SURGERY FOR GENERAL PRACTITIONERS
AND STUDENTS OF MEDICINE. By ASHLEY W. BARRETT, M.B. Lond., M.R.C.S., L.D.S., Dental Surgeon to, and Lecturer on Dental Surgery and Pathology in the Medical School of, the London Hospital. Second Edition, with Illustrations, crown 8vo, 3s. 6d

BODILY DEFORMITIES AND THEIR TREATMENT: A
Handbook of Practical Orthopædics. By H. A. REEVES, F.R.C.S. Ed., Senior Assistant Surgeon and Teacher of Practical Surgery at the London Hospital. With numerous Illustrations, crown 8vo, 8s. 6d.

₊ *Further Volumes will be announced in due course.*

WILLIAM MARTINDALE, F.C.S.
AND
W. WYNN WESTCOTT, M.B. LOND.

THE EXTRA PHARMACOPŒIA with the additions introduced into the British Pharmacopœia 1885; and Medical References, and a Therapeutic Index of Diseases and Symptoms. Sixth Edition, limp roan, med. 24mo, 7s. 6d.

[*Now ready.*

A. STANFORD MORTON, M.B., F.R.C.S. ENG.
Senior Assistant Surgeon, Royal South London Ophthalmic Hospital.

REFRACTION OF THE EYE: Its Diagnosis, and the Correction of its Errors, with Chapter on Keratoscopy. Third Edition. Small 8vo, 3s.

C. W. MANSELL MOULLIN, M.A., M.D. OXON., F.R.C.S. ENG.
Assistant Surgeon and Senior Demonstrator of Anatomy at the London Hospital.

SPRAINS; THEIR CONSEQUENCES AND TREATMENT. Crown 8vo, 5s.

WILLIAM MURRELL, M.D., F.R.C.P.
Lecturer on Materia Medica and Therapeutics at Westminster Hospital.

I.

MASSOTHERAPEUTICS; OR MASSAGE AS A MODE OF TREATMENT. Fifth Edition, crown 8vo, 4s. 6d.

[*Just ready.*

II.

WHAT TO DO IN CASES OF POISONING. Sixth Edition, royal 32mo, 3s. 6d.

[*Just published.*

G. OLIVER, M.D., F.R.C.P.

I.

ON BEDSIDE URINE TESTING: a Clinical Guide to the Observation of Urine in the course of Work. Fourth Edition, fcap. 8vo, 3s. 6d.

[*Just published.*

II.

THE HARROGATE WATERS: Data Chemical and Therapeutical, with notes on the Climate of Harrogate. Crown 8vo, with Map of the Wells, 3s. 6d.

K. W. OSTROM.

Instructor in Massage and Swedish Movements in the Hospital of the University of Pennsylvania.

MASSAGE AND THE ORIGINAL SWEDISH MOVEMENTS. With Illustrations, 12mo, 2s. 6d. *nett.*

[*Now ready.*

R. DOUGLAS POWELL, M.D., F.R.C.P., M.R.C.S.

Physician to the Hospital for Consumption and Diseases of the Chest at Brompton, Physician to the Middlesex Hospital.

DISEASES OF THE LUNGS AND PLEURÆ INCLUDING CONSUMPTION. Third Edition, with coloured plates and wood-engravings, 8vo, 16s.

FRANCIS H. RANKIN, M.D.

President of the Newport Medical Society.

HYGIENE OF CHILDHOOD: Suggestions for the care of children after the period of infancy to the completion of puberty. Crown 8vo, 3s.

SAMUEL RIDEAL, D.SC. (LOND.), F.I.C., F.C.S., F.G.S.

Fellow of University College, London.

I.

PRACTICAL ORGANIC CHEMISTRY. The detection and properties of some of the more important organic compounds. 12mo, 2s. 6d.

II.

PRACTICAL CHEMISTRY FOR MEDICAL STUDENTS, Required at the First Examination of the Conjoint Examining Board in England. Fcap. 8vo, 2s.

E. A. RIDSDALE.

Associate of the Royal School of Mines.

COSMIC EVOLUTION: being Speculations on the Origin of our Environment. Fcap. 8vo, 3s.

SYDNEY RINGER, M.D., F.R.S.

Professor of the Principles and Practice of Medicine in University College; Physician to, and Professor of Clinical Medicine in, University College Hospital.

A HANDBOOK OF THERAPEUTICS. Twelfth Edition, revised, 8vo, 15s.

FREDERICK T. ROBERTS, M.D., B.SC., F.R.C.P.

Examiner in Medicine at the University of London; Professor of Therapeutics in University College; Physician to University College Hospital; Physician to the Brompton Consumption Hospital, &c.

I.

A HANDBOOK OF THE THEORY AND PRACTICE OF MEDICINE. Seventh Edition, with Illustrations, large 8vo, 21s.

II.

THE OFFICINAL MATERIA MEDICA. Second Edit., entirely rewritten in accordance with the latest British Pharmacopœia, fcap. 8vo, 7s. 6d.

ROBSON ROOSE, M.D., F.R.C.P. EDIN.

I.

LEPROSY, AND ITS TREATMENT AS ILLUS- TRATED BY NORWEGIAN EXPERIENCE. Crown 8vo, 3s. 6d.

II.

GOUT, AND ITS RELATIONS TO DISEASES OF THE LIVER AND KIDNEYS. Sixth Edition, crown 8vo, 3s. 6d.

III.

NERVE PROSTRATION AND OTHER FUNC- TIONAL DISORDERS OF DAILY LIFE. Crown 8vo, 10s. 6d.

BERNARD ROTH, F.R.C.S.
Fellow of the Medical Society of London.

THE TREATMENT OF LATERAL CURVATURE OF THE SPINE. Demy 8vo, with Photographic and other Illustrations, 5s.

DR. B. S. SCHULTZE.

THE PATHOLOGY AND TREATMENT OF DIS- PLACEMENTS OF THE UTERUS. Translated by J. J. MACAN, M.A., M.R.C.S., and edited by A. V. MACAN B.A., M.B., Master of the Rotunda Lying-in Hospital, Dublin. With Illustrations, medium 8vo, 12s. 6d.

WM. JAPP SINCLAIR, M.A., M.D.
Honorary Physician to the Manchester Southern Hospital for Women and Children, and Manchester Maternity Hospital.

ON GONORRHŒAL INFECTION IN WOMEN. Post 8vo, 4s.

ALEXANDER J. C. SKENE, M.D.
Professor of Gynæcology in the Long Island College Hospital, Brooklyn.

TREATISE ON THE DISEASES OF WOMEN. With 251 engravings and 9 chromo-lithographs, medium 8vo, 28s.

ALDER SMITH, M.B. LOND., F.R.C.S.
Resident Medical Officer, Christ's Hospital, London.

RINGWORM: ITS DIAGNOSIS AND TREATMENT. Third Edition, rewritten and enlarged, with Illustrations, fcap. 8vo, 5s. 6d.

JOHN KENT SPENDER, M.D. LOND.
Physician to the Royal Mineral Water Hospital, Bath.

THE EARLY SYMPTOMS AND THE EARLY TREATMENT OF OSTEO-ARTHRITIS, commonly called Rheumatoid Arthritis. With special reference to the Bath Thermal Waters. Small 8vo, 2s. 6d.

LOUIS STARR, M.D.
Clinical Professor of Diseases of Children in the Hospital of the University of Pennsylvania.

HYGIENE OF THE NURSERY. Including the General Regimen and Feeding of Infants and Children, and the Domestic Management of the Ordinary Emergencies of Early Life. Second edition, with illustrations, crown 8vo, 3s. 6d.

W. R. H. STEWART, F.R.C.S., L.R.C.P. EDIN.
Aural Surgeon to the Great Northern Central Hospital; Surgeon to the London Throat Hospital, &c.

EPITOME OF DISEASES AND INJURIES OF THE EAR. Royal 32mo, 2s. 6d.

ADOLF STRÜMPELL, M.D.
Director of the Medical Clinic in the University of Erlangen.

A TEXT-BOOK OF MEDICINE FOR STUDENTS AND PRACTITIONERS. Translated from the latest German Edition by Dr. H. F. VICKERY and Dr. P. C. KNAPP, with Editorial Notes by Dr. F. C. SHATTUCK, Visiting Physician to the Massachusetts General Hospital, etc. Complete in one volume, with 111 Illustrations imp. 8vo, 28s.

LEWIS A. STIMSON, B.A., M.D.
Professor of Clinical Surgery in the Medical Faculty of the University of the City of New York, etc.

A MANUAL OF OPERATIVE SURGERY. With three hundred and forty-two Illustrations. Second Edition, post 8vo, 10s. 6d.

JUKES DE STYRAP, M.K.Q.C.P.
Physician-Extraordinary, late Physician in Ordinary to the Salop Infirmary; Consulting Physician to the South Salop and Montgomeryshire Infirmaries, etc.

I.
THE YOUNG PRACTITIONER: With practical hints and instructive suggestions, as subsidiary aids, for his guidance on entering into private practice. Demy 8vo, 7s. 6d. *nett.*

II.
A CODE OF MEDICAL ETHICS: With general and special rules for the guidance of the faculty and the public in the complex relations of professional life. Third edition, demy 8vo, 3s. *nett.*

III.
MEDICO-CHIRURGICAL TARIFFS. Fourth edition, revised and enlarged, fcap. 4to, 2s. *nett.*

IV.
THE YOUNG PRACTITIONER: HIS CODE AND TARIFF. Being the above three works in one volume. Demy 8vo, 10s. 6d. *nett.*

C. W. SUCKLING, M.D. LOND., M.R.C.P.
Professor of Materia Medica and Therapeutics at the Queen's College, Physician to the Queen's Hospital, Birmingham, etc.

ON THE DIAGNOSIS OF DISEASES OF THE BRAIN, SPINAL CORD, AND NERVES. With Illustrations, crown 8vo, 8s. 6d.

JOHN BLAND SUTTON, F.R.C.S.
Lecturer on Comparative Anatomy, and Assistant Surgeon to the Middlesex Hospital.

LIGAMENTS: THEIR NATURE AND MORPHOLOGY. Wood engravings, post 8vo, 4s. 6d.

EUGENE S. TALBOT, M.D., D.D.S.
Professor of Dental Surgery in the Women's Medical College.

IRREGULARITIES OF THE TEETH AND THEIR TREATMENT. With 152 Illustrations, royal 8vo, 10s. 6d.

E. G. WHITTLE, M.D. LOND., F.R.C.S. ENG.
Senior Surgeon to the Royal Alexandra Hospital, for Sick Children, Brighton.

CONGESTIVE NEURASTHENIA, OR INSOMNIA AND NERVE DEPRESSION. Crown 8vo, 3s. 6d.

JOHN WILLIAMS, M.D., F.R.C.P.
Professor of Midwifery in University College, London ; Obstetric Physician to University College Hospital

CANCER OF THE UTERUS: BEING THE HARVEIAN LECTURES FOR 1886. Illustrated with Lithographic Plates, royal 8vo, 10s. 6d.

BERTRAM C. A. WINDLE, M.A., M.D. DUBL.
Professor of Anatomy in the Queen's College, Birmingham ; Examiner in Anatomy in the Universities of Cambridge and Durham.

A HANDBOOK OF SURFACE ANATOMY AND LANDMARKS. Illustrations, post 8vo, 3s. 6d.
[*Just published.*]

EDWARD WOAKES, M.D.
Senior Aural Surgeon and Lecturer on Aural Surgery at the London Hospital, Surgeon to the London Throat Hospital.

I.
POST-NASAL CATARRH, AND DISEASES OF THE NOSE CAUSING DEAFNESS. With Illustrations, crown 8vo, 6s. 6d.

II.
NASAL POLYPUS: WITH NEURALGIA, HAY-FEVER, AND ASTHMA, IN RELATION TO ETH-MOIDITIS. With Illustrations, crown 8vo, 4s. 6d.

DAVID YOUNG, M.C., M.B., M.D.
Fellow of, and late Examiner in Midwifery to, the University of Bombay, etc.

ROME IN WINTER AND THE TUSCAN HILLS IN SUMMER. A Contribution to the Climate of Italy. Small 8vo, 6s.

CLINICAL CHARTS FOR TEMPERATURE OBSERVATIONS, ETC.
Arranged by W. RIGDEN, M.R.C.S.   Price 1s. per doz., 7s. per 100, 15s. per 250, 28s. per 500, 50s. per 1000.
Each Chart is arranged for four weeks, and is ruled at the back for making notes of cases; they are convenient in size, and are suitable both for hospital and private practice

PERIODICAL WORKS PUBLISHED BY H. K. LEWIS.

THE NEW SYDENHAM SOCIETY'S PUBLI-CATIONS. Annual Subscription, One Guinea.

Report of the Society, with Complete List of Works and other information, gratis on application.

THE BRITISH JOURNAL OF DERMATOLOGY. Edited by Malcolm Morris and H. G. Brooke. Published monthly, 1s. per no. Annual Subscription, 12s., post free.

THE NEW YORK MEDICAL JOURNAL. A Weekly Review of Medicine. Annual Subscription, One Guinea, post free.

THE THERAPEUTIC GAZETTE. A Monthly Journal, devoted to the Science of Pharmacology, and to the introduction of New Therapeutic Agents. Edited by Drs. H. C. Wood and R. M. Smith. Annual Subscription, 10s., post free.

THE GLASGOW MEDICAL JOURNAL. Published Monthly. Annual Subscription, 20s., post free. Single numbers, 2s. each.

LIVERPOOL MEDICO-CHIRURGICAL JOURNAL, including the Proceedings of the Liverpool Medical Institution. Published twice yearly, 3s. 6d. each number.

MIDDLESEX HOSPITAL. Reports of the Medical, Surgical, and Pathological Registrars for 1883 to 1888. Demy 8vo, 2s. 6d. *nett* each volume.

TRANSACTIONS OF THE COLLEGE OF PHYSICIANS OF PHILA-DELPHIA. Volumes I. to VI., 8vo, 10s. 6d. each.

*** Mr. Lewis is in constant communication with the leading publishing firms in America and has transactions with them for the sale of his publications in that country. Advantageous arrangements are made in the interests of Authors for the publishing of their works in the United States.

Mr. Lewis's publications can be procured of any Bookseller in any part of the world.

Complete Catalogue of Publications post free on application.

Printed by H K. Lewis, Gower Street, London, W.C.

www.ingramcontent.com/pod-product-compliance
Lightning Source LLC
Chambersburg PA
CBHW021955190326
41519CB00009B/1271